Current Natural Sciences

AF328867

Thérèse ENCRENAZ,
James LEQUEUX
and Fabienne CASOLI

Planets and Life

Science Press

edp sciences

Printed in France

EDP Sciences – ISBN(print): 978-2-7598-2563-9 – ISBN(ebook): 978-2-7598-2570-7
DOI: 10.1051/978-2-7598-2563-9

Foreword

With a surface pressure of 1 bar and an average temperature of 15°C, the Earth is the only planet in the Solar System with liquid water on its surface. The surface conditions of Venus and Mars are extremely different, with a pressure close to one hundred times the Earth's value on Venus and less than one hundredth on Mars, and a temperature ranging from more than 460°C on Venus to about –50°C on Mars. How could these three planets, starting from relatively comparable initial conditions, have evolved to the extreme diversity we observe today? Understanding the origin and evolution of the atmospheres of the three terrestrial planets – Venus, Earth and Mars – is a major challenge for planetology. Highlighting the physical or chemical factors that were and are still at play appears as a first step to better understand the context in which life appeared and developed on the Earth.

This question takes a new dimension with the discovery, since a quarter of century, of thousands of extrasolar planets, among which are many "rocky" ones, *i.e.*, with a surface like the terrestrial planets of the Solar System. They are called, according to their mass, "exo-Earths" or "super-Earths". Could some of these exoplanets harbor life? With their discovery, the question "Are we alone in the Universe?", which is as old as humanity itself, is no longer confined to our Solar System, and the field of possibilities opens up to infinity. In this new context, it is more than ever necessary to understand the evolution of the rocky planets and to identify the factors that determine their habitability, *i.e.*, their capacity to allow the emergence and development of life. These factors can be multiple. Some are of physico-chemical nature (pressure and temperature, atmospheric composition); others are related to the planet's environment (nature of the star, presence of a magnetosphere) or some of its parameters (ellipticity of the orbit, obliquity of the planet axis, rotation period).

For more than two millennia, the quest for extraterrestrial life, present from the earliest ages of mankind, has been based on philosophical considerations. It is only since the end of the 19th century that astronomers were able to begin to approach the question in a scientific manner, first with the observation of the planets that surround us, and then, half a century later, with the search for exoplanets around other stars. The end of the 20th century witnessed an avalanche of discoveries that continued and amplified until now. Increasingly complex planetary space missions explore the soil and subsoil of Mars for possible traces of fossil life; others, in the coming decades, will explore the satellites of the giant planets of the outer Solar

DOI: 10.1051/978-2-7598-2563-9.c901

System, some of which may harbor an ocean of liquid water beneath their icy surfaces. In parallel, we now have the possibility to determine the nature of the exoplanets and, in some cases, their atmospheric composition. Among the rocky exoplanets known today, several tens could have a temperature compatible with the presence of liquid water. In one or two decades, this research will be refined to allow, perhaps, to discover on one or more of them oxygen or its derivative, ozone, a possible signature of the presence of life.

In this context of abundant and constantly evolving research, it seemed useful to try to better define the criteria for habitability of rocky exoplanets, those that could shelter life. This book is to some extent the continuation of the book "The Exoplanet Revolution", by J. Lequeux, T. Encrenaz and F. Casoli, published in the same collection in 2020. Like it, it is addressed to all audiences interested in astronomy, planetology and the search for extraterrestrial life. Here, we start from the planets we know well, the three terrestrial planets of the Solar System possessing an atmosphere, to analyze the various physico-chemical mechanisms that could have been responsible for their divergent evolution. Then we will try to extrapolate these results to the rocky extrasolar planets, in order to understand the possible mechanisms of their evolution and to apprehend what could be their conditions of habitability. Finally, we conclude this work by an analysis of the means that could allow us to discover possible traces of life, or even to communicate with eventual distant civilizations.

Contents

Foreword III

CHAPTER 1

Introduction 1

CHAPTER 2

The Formation of Terrestrial Planets 9
2.1   From Antiquity, the Myth of the Plurality of Worlds. 9
2.2   The Primordial Nebula Model. 11
2.3   The Formation of Stars and Discs 13
2.4   The Formation of Terrestrial and Giant Planets 16
2.5   The Migration of Planets 19
2.6   The Late Heavy Bombardment and Its Consequences 21
2.7   The Formation of Planets in Exoplanetary Systems. 23
2.8   The Primary Atmospheres of the Terrestrial Planets 24
2.9   What Atmospheres for Rocky Exoplanets? 25

CHAPTER 3

The Exploration of Terrestrial Planets 27
3.1   The First Modern Observations 27
3.2   The Myth of the Martian Canals 29
3.3   The Physical Nature of Planets 30
3.4   The Beginning of the Space Era 30
3.5   The Viking Mission: Hopes and Disillusions 32
3.6   From Mars to Venus... 33
3.7   The Renewal of Martian Exploration 35
3.8   Return to Venus 39
3.9   Mars and Venus Today 41
3.10   Between Venus and Mars, the Earth 42
3.11   Towards a Comparative Study of the Terrestrial Planets 46

CHAPTER 4

Venus, Earth and Mars: A Diverging Evolution 47
4.1   The Astonishing Variety of Terrestrial Planets 49
4.2   And Yet... Common Characteristics 52
       4.2.1   The Thermal Structure of the Terrestrial Planets 53
       4.2.2   Atmospheric Circulation................................. 54
       4.2.3   Internal Structure and Volcanism 55
4.3   Terrestrial Planets at the Origin 56
       4.3.1   Secondary Atmospheres 56
       4.3.2   Primitive Atmospheres Rich in Water...................... 57
       4.3.3   The Paradox of the "Young Sun" 59
4.4   History of the Terrestrial Planets: A Divergent Evolution 61
       4.4.1   Venus: The Ravages of a Runaway Greenhouse Effect........ 61
       4.4.2   Mars: A Planet on the Verge of Geological Extinction 62
       4.4.3   The Earth, Ideally Located in Relation to the Sun 63

CHAPTER 5

The Appearance of Life 65
5.1   What is Life? 65
5.2   From Spontaneous Generation to Primordial Soup 66
5.3   The First Experiments in Prebiotic Chemistry 68
5.4   The Building Blocks of Terrestrial Life 70
5.5   The Origin of Prebiotic Molecules 73
5.6   The Rise of Complexity from Prebiotic Molecules 75
5.7   The Formation of Cells 76
5.8   Metabolism and the Question of Energy 78
5.9   The Genetic Code.. 79
5.10   The Ancestor of All Living Beings?............................... 79
5.11   Life on Earth as a Model for Life on Other Planets? 81
5.12   The Beginnings of Life on Earth 82
5.13   Life on Exoplanets 83

CHAPTER 6

The Development of Life on Earth 85
6.1   The Paradox of the "Young Sun" 85
6.2   The Major Stages in the Evolution of the Earth's Climate 87
       6.2.1   From Hadean to Archean 87
       6.2.2   From Archean to Proterozoic: The Great Oxidation Event 89
       6.2.3   The Phanerozoic: Life on the Continents 91
6.3   What Future for the Earth's Atmosphere? 94
6.4   What Lessons for Exobiology? 96

CHAPTER 7

Life in the Solar System? 99
7.1 The Habitability Zone in the Solar System 99
7.2 A Past Ocean on Venus? 102
7.3 Searching for Traces of Life on Mars 104
7.4 Other Niches in the Solar System 106

CHAPTER 8

How to Search for Life on Rocky Exoplanets? 113
8.1 The Discovery of Exoplanets: Where Do We Stand? 113
8.2 The Exoplanet Concept: An Old Idea 117
8.3 Early Discoveries 118
8.4 The Successes of Velocimetry 119
8.5 A New Step: The Transit Method 120
8.6 How to Search for Life on an Exoplanet? 124
8.7 Satellites Around Giant Exoplanets? 126
8.8 How to Determine the Atmospheric Composition of an Exoplanet? 127
8.9 How to Search for Life from the Spectrum of an Exoplanet? 129

CHAPTER 9

Conclusions: Some Future Directions in Exobiology 135
9.1 The Future of Mars Exploration 135
9.2 How to Detect Traces of Life *In Situ?* 139
9.3 Towards an Inhabited Exploration of Mars? 139
9.4 Towards External Satellites, Other Possible Niches for Life 141
9.5 Exploring Exoplanets: The Prospects 142
9.6 What If We Were Not Alone? 146

Glossary 149
Bibliography 155

Chapter 1

Introduction

Within the planets of the Solar System, our closest neighbors, Venus and Mars, are undoubtedly those that surprise us the most. While they belong, like the Earth, to the family of rocky or "terrestrial" planets, and are, like the Earth, with an atmosphere, they have radically different surface conditions: on Venus, the pressure is about a hundred times the Earth's atmospheric pressure and the temperature reaches 460°C, while on Mars, the surface pressure is less than a hundredth of a bar, and the average temperature is around −50°C! Between these two extremes, the Earth holds an intermediate position, with a surface pressure of the order of one bar and an average surface temperature of 15°C (figure 1.1). How could these three rocky planets, all formed in the inner Solar System 4.5 billion years ago, from relatively similar initial conditions, have evolved towards such radically divergent fates? This question, one of the most fundamental of today's planetology, is addressed in this book.

Understanding the comparative evolution of the terrestrial planets of the Solar System is very important to decipher the origin and evolution of our own environment. With the discovery, over the last two decades, of thousands of extrasolar planets in orbit around neighboring stars, including a growing number of rocky exoplanets, the debate takes on a new dimension. Indeed, the major question about rocky exoplanets is that of their potential habitability: if there is an extraterrestrial form of life, is not this new class of objects the best place to look for it? The first step in this search is to determine the temperature and pressure of their atmosphere. An astronomer observing the planets of the inner Solar System from the nearest star, Proxima Centauri, would find it hard to imagine the extreme diversity of conditions on the surfaces of the Earth's planets. That is to say that the physical properties of rocky exoplanets undoubtedly hold many surprises in store for us, and if we cannot study them in detail today, we must understand the mechanisms that are responsible, in the Solar System, for the divergent evolution of the terrestrial planets.

Why does the planet Venus have such a high temperature today? We now know the reason, and there is a lot of talk these days about it concerning the global warming of our own planet: it is the greenhouse effect (figure 1.2). What is it all

DOI: 10.1051/978-2-7598-2563-9.c001

F$_{\text{IG}}$. 1.1 – Venus, Earth and Mars: relatively similar initial conditions but divergent fates. The relative dimensions of the three planets are respected. © NASA.

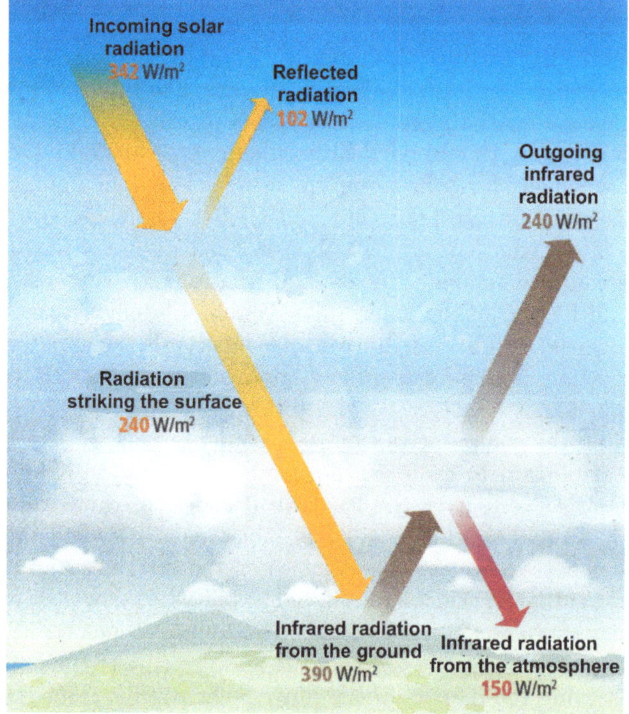

F$_{\text{IG}}$. 1.2 – The greenhouse effect (average figures for the Earth). The atmosphere is transparent in the visible range and solar radiation reaches the surface. When heated, the surface emits infrared radiation which is absorbed by the atmosphere if it contains certain gases with intense infrared vibration modes, such as carbon dioxide CO_2, methane CH_4 or water vapor H_2O. In the case of Venus, CO_2 and H_2O (in the planet's past) have been responsible for the runaway greenhouse effect during its history.

about? The greenhouse effect occurs when an atmosphere is transparent to visible radiation, but opaque to infrared radiation. The glass walls of a greenhouse allow visible solar radiation to pass through, causing the interior of the greenhouse to heat up; as glass absorbs the infrared radiation emitted from the inside, the temperature rises and the effect is amplified. In the case of a global atmosphere, the greenhouse effect occurs if atmospheric gases absorb infrared radiation; this is the case of carbon dioxide CO_2 and also of water vapor H_2O. In the case of the Earth, the main constituents, nitrogen N_2 and oxygen O_2 do not have vibration modes in the infrared, and therefore do not contribute to the greenhouse effect; the increasing emissions of CO_2 are the primary cause of global warming; other gases, such as water H_2O and methane CH_4, also contribute to the greenhouse effect. In the case of Venus, the situation is different. As for Mars, the dominant gas is carbon dioxide, with a small proportion (a few percent) of nitrogen. Since, moreover, the surface pressure of Venus is very large, the very high surface temperature is the result of a strong greenhouse effect due to CO_2.

Why is the Earth's atmospheric composition (about four-fifths molecular nitrogen and one-fifth oxygen) so different from that of Venus and Mars? This is where another key molecule comes into play: water H_2O. It is now established (we will see later how) that the primitive atmospheres of the three planets, Venus, Earth and Mars, were globally similar, with large quantities of carbon dioxide and water and a small proportion of molecular nitrogen. In the case of Venus, closer to the Sun than the Earth, water was, at some point in its history, in the form of vapor, thus contributing to the planet's runaway greenhouse effect. On the other hand, the distance from the Earth to the Sun is such that the Earth's water was found in liquid form in the oceans; carbon dioxide, also very abundant, was trapped at the bottom of the oceans in the form of limestone; thus the two main greenhouse gases disappeared to a large extent from the Earth's atmosphere, allowing the planet to maintain temperate conditions throughout its history.

How can we explain the evolution of Mars? The planet has two notable differences with respect to Venus and Earth: being more distant from the Sun, it is colder (today, water cannot stay on the surface in liquid form) but it is also smaller: its mass is only one tenth of that of the Earth. With a gravity field much lower than that of Venus and the Earth, it could not capture, like its neighbors, a thick atmosphere; it is thought that its primitive atmosphere (today largely disappeared by escape) could not have gone beyond a pressure of a bar. Because of its smaller volume, its internal energy, mainly due to the disintegration of the radioactive elements that the planet contains, was also much less than that of its neighbors, resulting in reduced volcanic and tectonic activity that eventually died out over the course of history.

If the main lines of the evolution of the terrestrial planets seem well defined to us, many open questions remain, starting with that of their habitability. Did Mars and Venus ever shelter a form of life? We are today quite unable to answer. The question becomes even more complicated if we take into account the amount of solar radiation at the beginning of the history of the planets: it is the paradox of the "young Sun". Models of stellar evolution tell us that, about four billion years ago, the Sun's radiation in visible light (which corresponds to the maximum of its energy) was only

70% of its present value. The equilibrium temperatures of the surfaces of our three terrestrial planets were thus lower than what they are today. Hence a major consequence for Venus: the temperature may have been compatible with that of liquid water, and the primitive Venus may have been covered with oceans, and perhaps even, who knows, sheltered life! Unfortunately, if these conditions existed, they did not last: as solar radiation increased, water vaporized (contributing for a time to the greenhouse effect), then was dissociated by solar ultraviolet radiation into hydrogen and oxygen atoms that escaped to the outside. If an ocean (and a fortiori life) ever existed at the beginning of the history of Venus, we will probably never know anything about it, because the traces of it have irremediably disappeared: the surface of Venus is indeed covered with relatively recent volcanoes, less than a billion years old.

The paradox of the "young Sun" also raises questions that are still poorly resolved in the case of other terrestrial planets. How can we explain that the Earth, at the beginning of its history, escaped a "global snowball" episode, its equilibrium temperature being too low to be compatible with the presence of liquid water? One possible hypothesis is that of volcanic eruptions that released into the atmosphere enough greenhouse gases (CO_2 but also CH_4). The same question arises even more acutely in the case of planet Mars. As we will see further on, many clues testify to the presence of liquid water on the surface in the distant past of this planet. How could liquid water have remained on Mars when its equilibrium temperature was incompatible (by several tens of degrees!) with the presence of liquid water? The question is still open.

With the Earth and Mars, we fortunately have an avenue of research: unlike Venus, these two planets keep archives on their surfaces that allow us to trace their history back nearly four billion years. This is why Mars is still the subject of sustained space exploration, with the objective of searching for possible traces of a past – or even present – life. Failing that, research is focused on the characterization of "habitable" sites, *i.e.* sites that meet the physico-chemical criteria compatible with the emergence of life (figure 1.3). These criteria relate in particular to soil acidity (preferably neutral), its salinity (moderate) and its chemical composition (including elements C, H, N, O, P, S). Can we go further? Space exploration of Mars continues and the future will tell us...

From terrestrial planets to rocky exoplanets, it is only a step away. It has been more than twenty years since planets, known as "extrasolar planets" or "exoplanets", were discovered around stars close to the Sun. To everyone's surprise, the first objects discovered (the easiest to detect from an observational point of view), were giant exoplanets very close to their star! This discovery created a real conceptual revolution, challenging our own understanding of the Solar System. Indeed, according to the formation model of the Solar System, widely accepted today in the scientific community, giant planets were formed far from the Sun, by accumulation of gas around an icy core, whereas terrestrial planets were formed from a smaller and denser rocky core. The first discoveries of exoplanets thus showed that the model of the Solar System was not universal... The explanation of this paradox originates in the motion of the planets within the protoplanetary disc: we speak of "migration", a process that, until then, was not much taken into account by planetary scientists.

Fig. 1.3 – The Mars Science Laboratory (MSL) space mission, with its rover "Curiosity", launched by NASA in November 2011 and in operation on the surface of Mars since August 2012, has the main objective of determining whether conditions favorable to life may have existed on Mars, in particular through the search for organic molecules. The rover has indeed identified, near Mount Sharp, the remains of an ancient lake that could constitute a "habitable" environment. © NASA.

Very effective within exoplanetary systems, this process has also proved to be important for understanding the dynamic history of our own Solar System.

While the first exoplanets detected were mostly giant planets, identified from the Earth from the oscillations of their host star with respect to the planet's motion, a new revolution occurred with the launch of the *CoRoT* and, even more, the *Kepler* space missions, dedicated to the detection of exoplanets during their passage (called transit) in front of their host star. *Kepler* has thus detected thousands of new exoplanets, including objects of all sizes, among which are new families of planets: "mini-Neptunes", "super-Earths", even "exo-Earths". Although the physical nature of these planets is not yet known, it is very likely that many rocky exoplanets are among the new discoveries. The first measurements of the atmospheres of the exoplanets show us that water vapor is often present there. Among the rocky exoplanets, some of them, located at the right distance from their host star – in what is called the "habitability zone" – could harbor water in liquid form, and thus perhaps

FIG. 1.4 – Among the detected rocky exoplanets, some could experience a temperate environment, given the amount of light they receive from their host star: they are in the "habitable zone" of their star, the one whose temperature is compatible with the presence of liquid water on their surface. However, the composition of their atmospheres remains to be identified, in order to know if water is present, a fortiori in liquid form. This figure represents a sample selected by researchers at Arecibo in 2020. Only the sizes of the exoplanets are meaningful.

constitute potential niches for the emergence and development of life (figure 1.4). The exploration of exoplanets thus opens up immense perspectives in terms of exobiology.

A first question is obvious: if life ever existed on the surface of an exoplanet, how could we identify it? We will come back to this notion in the course of this book. Let us simply say that, according to biologists, living matter can be characterized by the capacity for reproduction, the ability to use the energy of the environment, the separation from this environment, the creation of an organization, and finally the capacity for evolution by mutation. Of course, the forms that life can take can be infinitely diverse, as shown by the only example we know of, that of the Earth. We still do not know how life appeared on Earth. However, as we will see later, chemists and biologists, again based on our experience on Earth, agree to define some essential conditions: the presence of carbon, liquid water, a source of energy, and sustainability. These are the criteria that will be used in the quest for extraterrestrial life, whether within the Solar System or beyond.

How can the presence of life be demonstrated by observing the atmosphere of an exoplanet? Given the immense distances that make illusory the prospect of a space mission, even a robotic one, to approach the object in question, we are limited to remote sensing methods, the most promising of which is spectroscopic characterization. We will have to determine which constituents, in the atmosphere or on the surface of the object, could betray the presence of life. Some tracks are already opening up: in the atmosphere, the presence of molecular oxygen O_2, in substantial quantities, as well as its photochemical derivative, ozone O_3, seem to be a fairly convincing clue, but we will see that the reality is more complex and that the simultaneous presence of other constituents – H_2O, CH_4, CO_2, N_2O... – is probably also necessary for the characterization of life. On the surface, the presence of chlorophyll would be a determining diagnosis, but we will see that its spectroscopic detection is almost impossible. And let us not forget that life on another planet may have taken very different forms from the photosynthesis we know on Earth.

The main goal of this book is to explore the habitability conditions of rocky exoplanets, starting from what we know: the terrestrial planets of the Solar System with an atmosphere. By observing their divergent evolution, by studying – as far as possible – their past or present habitability conditions, we will try to extrapolate these notions to the rocky exoplanets, of which we know very little at the moment. Among the currently known exoplanets, we will try to identify the most favorable candidates, those which seem to be the best placed in relation to the habitability zone of their star. Finally, we will try to define which observations would allow us to conclude to the possible or probable presence of life. Our first objective is the search for life forms on the surface of an exoplanet, likely to present similarities with the development of life on Earth over the last 600 million years. There are other niches potentially favorable to life: these are the oceans of liquid water that are sheltered under the surface of several of the outer satellites of the Solar System; this is the case, in particular, of Europa, a satellite of Jupiter, and Enceladus, a satellite of Saturn. Such environments could exist around possible satellites in orbit around giant exoplanets; although, with a few exceptions, these have yet to be discovered, their existence is very plausible according to models of planetary formation. However, the detection of life forms within these aqueous media seems *a priori* much more difficult than for those which would have developed on the surface of an exoplanet.

In the first part of this book, we describe the past history of the terrestrial planets, starting from the model of the formation of the Solar System which will make us discover the emergence of two classes of planets, the terrestrial planets near the Sun and the giant planets beyond. This model will allow us to understand the nature of the gases originally present in the atmospheres of these planets. We will also see how the migration of the giant planets during their history has influenced the dynamic history of the whole Solar System. In a second part, we will describe the three terrestrial planets with atmospheres, the history of their exploration, the state of our knowledge about them and finally their different evolution scenarios. Finally, in the third part, we will extrapolate this knowledge to what we know about rocky exoplanets. We will look for targets potentially favorable to the search for life and we will try to define the observations that could allow us, within one or several decades, to finally discover, who knows, the signs of extraterrestrial life. To conclude, we will present some lines of thought to try to define the future stages of exobiology, and we will address the question of a possible communication with extraterrestrial civilizations, if we manage to discover their existence.

Chapter 2

The Formation of Terrestrial Planets

2.1 From Antiquity, the Myth of the Plurality of Worlds

From the eve of civilizations, man has never ceased to question himself about his place in the Universe. Since Antiquity, Greek philosophers have raised this question, with a variety of answers. Plato (428–348 B.C.), a proponent of the uniqueness of the Earth, wrote in *Timæus*:

> In order that this World [the Earth] may be like, in its unity, a perfect creature, its Author did not make two of them, nor an infinite number, but this Heaven was born, remains and will be forever one and unique.

However, Epicurus (ca. 342–270 B.C.), a century later, supported the opposite concept of an infinite number of worlds. Supporter, like Democritus (ca. 460–370 B.C.) before him, of the "atomist" thesis according to which matter consists of invisible atoms that can combine in various forms, he wrote in his *Letter to Herodotus*:

> It is not only the number of atoms, but the number of worlds that is infinite in the Universe. There is an infinite number of worlds similar to ours and an infinite number of different worlds.

At the same time, the astronomer and mathematician Aristarchus of Samos (370–230 B.C.), considering the diameters of the Sun and the Moon, stated for the first time the hypothesis of heliocentrism, which is also mentioned by Archimedes (287–212 B.C.) in the preface to his treatise *Arenarius*; but this hypothesis, in contradiction to the geocentric conception of Aristotle (384–322 B.C.), quickly felt into oblivion.

Three centuries later, the Latin philosopher Lucretius (ca. 98–55 B.C.) took up Epicurus' ideas in his poem *De Rerum Natura*:

> The sky, the Earth, the Sun, the Moon, the sea, all the bodies are not unique, but rather infinite in number.

DOI: 10.1051/978-2-7598-2563-9.c002
© Science Press, EDP Sciences, 2021

Epicurus, like Lucretia and their followers, were fiercely opposed to gods and religion. The development of Christianity plunged their theories into oblivion for nearly a millennium, until Nicolaus Cusanus (1401–1464) and the Copernican revolution. Then, many astronomers and philosophers took up the concept of the plurality of worlds; like their Greek predecessors, they most often associated the notion of habitability with it. The most emblematic among them is undoubtedly Giordano Bruno (1548–1600) who, in his works *La Cena de la Ceneri* (The Ashes Banquet) and *De l'Infinito Universo e Mondi*, published in 1584, postulated that the stars are other suns, no doubt surrounded by planets that could themselves be inhabited. Sentenced to be burned at the stake by the Church, Giordano Bruno paid for his audacity with his life.

If Giordano Bruno's convictions were mainly philosophical, his ideas were adopted to various extents by astronomers who, after Galileo, rallied to heliocentrism. Thus Johannes Kepler (1671–1630) published after his death a work of "science fiction", the *Somnium*, in which he envisaged a life on the Moon. The idea was taken up, in a philosophical and literary fashion, by Francis Goldwin (1562–1633) and John Wilkins (1614–1672) in England, then in France by Cyrano de Bergerac (1619–1655) and later by Voltaire (1694–1778). These ideas were also developed by Christiaan Huygens in his *Cosmotheoros* published in 1695, and by Bernard le Bovier de Fontenelle (1657–1757) in his *Entretiens sur la pluralité des mondes* (Conversations on the plurality of worlds) of 1686. In the preface to a reprint of this work, Jérôme de Lalande wrote in 1801:

> The resemblance is so great between the Earth and the other planets, that if one admits that the Earth was made to be inhabited, one cannot refuse to admit that the planets are also inhabited,

and further:

> What I say about the planets that revolve around the Sun will naturally extend to all the planetary systems that surround the stars.

A century later, these ideas were taken up anew and popularized by the astronomer Camille Flammarion (1842–1925) in his work *La pluralité des mondes habités* (The plurality of inhabited worlds) published in 1862, which led to his dismissal from the Paris Observatory by its director at the time, Urbain Le Verrier. In 1892, Flammarion joined the thesis of the "Martian canals" defended by Giovanni Schiaparelli (1835–1910) and Percival Lowell (1855–1916): see chapter 3. The myth of life on Mars continued until the advent of the space age.

But what about life outside the Solar System? The debate has continued among biologists who wonder about the origin of life. Thus Jacques Monod (1910–1976), in *Le Hasard et la Nécessité* (*Chance and Necessity*), published in 1970, expressed the conviction that man is alone in the Universe. But the search for extraterrestrial life went on, nevertheless, in multiple forms, in the Solar System and beyond, and attempts to communicate with possible extraterrestrial civilizations continue through the SETI project (see chapter 9).

2.2 The Primordial Nebula Model

Let us now return to the formation of the Solar System. Until the 17th century, knowledge was too rudimentary and minds too cluttered with metaphysical considerations to approach cosmogony, *i.e.* the theory of the formation of the Solar System, in a scientific manner. Even in the seventeenth century, when the laws governing the motion of the planets were known, there was no idea of the time scales involved, and little or nothing, at least before Newton, was known about the physics on which to base a coherent theory of the formation of the Sun and its planetary system.

However, René Descartes (1596–1650) was one of the first to try his hand at it, in his work *Le Monde*. This book was written from 1629 to 1633, but its author gave up publishing it during his lifetime for fear of hostile reactions on the part of the Church similar to those suffered by Galileo: it did not appear until 1664. For Descartes, the Universe is organized from the initial chaos according to the motions and encounters of great *tourbillons* (vortices) of matter: the Solar System was born in one of these vortices, and other planetary systems in others (figure 2.1). Giordano Bruno's heretical ideas on the plurality of worlds have been emulated. If Descartes' cosmogony did not rest on any reasonable physical basis, it had the merit of conceiving that the Universe can be in evolution, an evolution governed by relatively simple laws: its present state keeps traces of its origin and its history, a history that took

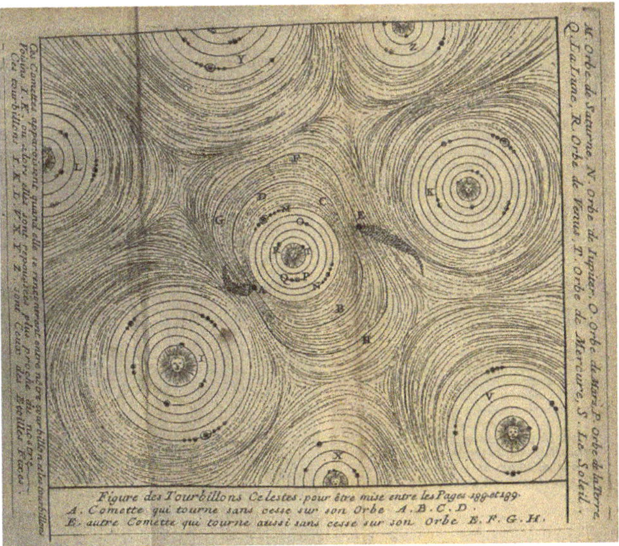

FIG. 2.1 – For Descartes, planetary systems were born in vortices. Descartes had only described the formation of the Solar System, but it is clear from reading him that he thought that planetary systems were formed around other stars. His successors, such as Nicolas Bion (1652–1733) from whom we borrow this 1702 engraving, represented them explicitly (© Bibliothèque de l'Observatoire de Paris).

place naturally and smoothly, without miracles or catastrophes. It is a remarkably modern conception for the time, prefiguring the evolutionary cosmogonic models to come. It can be contrasted with catastrophic cosmogonies such as the one that Georges-Louis Leclerc, Count of Buffon (1707–1788), set out in his *Theory of the Earth* of 1749 and then in *The Epochs of Nature* of 1778. For Buffon, the Sun and comets pre-existed the planets, which were born together from a catastrophe, the encounter of a comet with the Sun. The evolutionary models and the catastrophic models, which moreover were proposed before Buffon, were confronted for a long time on the basis of arguments that were often more ideological than scientific.

Immanuel Kant (1724–1804) adopted an attitude similar to that of Descartes. Knowledge in physics was then improved thanks to Newton, and his cosmogony is more elaborate than that of Descartes. In his *Allgemeine Naturgeschichte und Theorie der Himmels* (Universal history of nature and theory of the Sky) of 1755, he proposed that all the matter that forms the Solar System originally existed as a large nebulous mass; it would have contracted into a flat rotating disc under the effect of its own gravitation, then distributed between the Sun and the planets, born of inhomogeneities in the disc. The idea of the rotating disc obviously came from the fact that all the planets rotate in the same direction on orbits located in about the same plane. Kant's theory did not seem to him to be in any way contrary to religion, for he proclaimed that the laws of physics have no other reason than to tend to fulfill the purpose that divine wisdom has proposed to itself, and of which we see the result today.

Perhaps without knowing Kant, Pierre–Simon Laplace (1749–1813) put forward a very similar theory in his *Exposition du système du monde*, the first edition of which dates from 1796. In subsequent editions, he continued to refine it by incorporating new discoveries, notably that of the "small planets", the asteroids that orbit the Sun between the orbits of Mars and Jupiter. This last discovery, which filled a gap in the distribution of the distances of the planets from the Sun, seemed to him a confirmation of the idea that dominates his work: the current arrangement of the Solar System results from its origin, and owes nothing to chance. Laplace was familiar with the observations of William Herschel (1738–1822), who saw a central star in many nebulae. Both thought that these were primitive nebulae in which a sun had already formed, supporting Laplace's theory. Yet they were wrong: we now know that the vast majority of these objects, planetary nebulae, are made of a star at the end of its life that has ejected some of its matter in the form of a gas that it illuminates.

Laplace's authority and his popularizing talents were so great that his "nebular hypothesis" was to dominate the entire 19th century. However, several problems arose. The most serious, which seems to have been first brought to light in 1861 by Jacques Babinet (1794–1872), concerns the distribution of angular momentum in the Solar System. Indeed, it was expected that the angular momentum of rotation was preserved during the contraction of the initial cloud and was more or less evenly distributed between the various parts of the Solar System. The reality is very different: the Sun rotates slowly (one rotation in 27 days), and almost all the angular momentum is found in the revolution of the planets, especially Jupiter. Moreover, at the end of the 19th century, Henri Poincaré (1854–1912) began to doubt the long-term stability of the Solar System. Chance could perhaps play a role in the formation and evolution of this system, which tends to undermine the basic principles of Laplace's theory: moreover, at that time when the Darwinian theory of evolution was triumphing, why would the Solar System not evolve?

The problem of angular momentum is at the origin of a renewal of catastrophic cosmogony: the astronomer Forest Ray Moulton (1872–1952) and the geologist Thomas Chamberlain (1843–1928) estimated in 1906 that the orbital angular momentum of the planets could only come from outside, and that the planetary system must originate from matter from the Sun torn by the attraction of another star passing nearby, which would have communicated to this matter the observed angular momentum. This idea, an avatar of Buffon's, had much popular success in various forms at the time. An obvious consequence of this catastrophe theory is that, if it is true, planetary systems like ours must be extremely rare since the passage of a star very close to another can only be exceptional, given the immense distances between stars. But we know today that planetary systems are very frequent. In any case, the catastrophe theory did not wait for the discovery of so many exoplanets to collapse, when it was subjected to fine analyses, in particular that of Lyman Spitzer (1914–1997) in 1939: one had to return to Laplace and even, in a much modified form, to Descartes' vortices.

These vortices were reintroduced more or less simultaneously by several renowned astronomers, particularly in 1944 by Carl Friedrich von Weiszäcker (1912–2007), the one who, together with Hans Bethe (1906–2005), had found the origin of the energy of stars. He imagined that in the rotating gaseous disc resulting from the collapse of the initial cloud, vortices are established, whose system was reminiscent of concentric ball bearings. The planets would not form within the vortices as Descartes had imagined, but in the spaces between them. This theory did not resist criticism: in 1948, its author himself acknowledged that it had to be abandoned. His colleagues did the same at about the same time. Nevertheless, their work, as well as that of Gerard Kuiper (1905–1973) and others, had the merit of posing the problems well and paving the way for further research.

2.3 The Formation of Stars and Discs

We know today that stars and the planets around them form within interstellar clouds of gas and dust. Observation shows that the coldest clouds are the place of formation of small-mass stars. These stars are not detectable in visible light, but they can be observed in the infrared inside the cloud: if the dust contained in these clouds makes them opaque to light, it affects much less the radiation of longer wavelengths (figure 2.2). Newly formed stars appear at the periphery of the cloud when it dissipates under the effect of their radiation and the wind they eject. The giant interstellar clouds, which are warmer, are places where stars of all masses form (figure 2.3). The formation of massive stars appears to be a contagious and self-sustaining phenomenon: the pressure and shock waves generated by stellar winds and by the final explosion of stars of previous generations trigger the formation of new stars in what remains of the cloud or in nearby clouds.

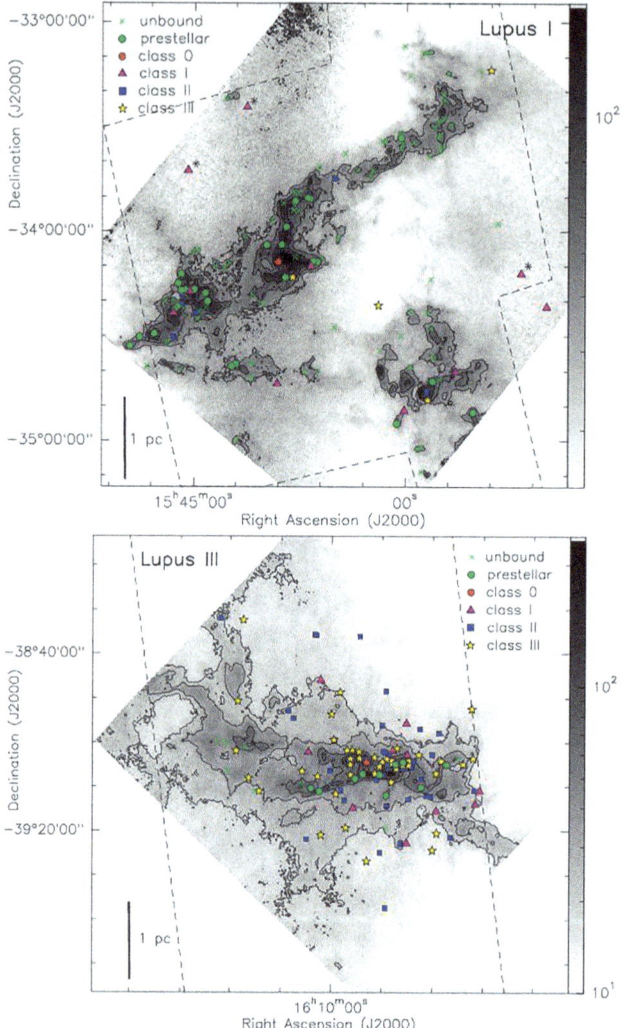

FIG. 2.2 – Star formation in two molecular clouds in the direction of the constellation Lupus, observed in the infrared with the European satellite Herschel. The column density of the gas is in gray; notice its filamentary structure. The stars in formation are represented by different symbols, from the protostellar condensations (green circles) to the stars having dissipated their protostellar disc, the planets being entirely formed (Class III, yellow stars). The intermediate stages are indicated by red circles (Class 0), purple triangles (Class I) and blue squares (Class II). See figure 2.4 for the definition of these classes. Star formation is much more advanced in Lupus III than in Lupus I. From Rygl *et al.*, *Astronomy & Astrophysics* 549, L1 (2013).

FIG. 2.3 – Star formation in the interstellar clouds associated with the Orion Nebula, as revealed by infrared observations. This false color image was obtained with one of the four 8-m diameter telescopes of the Very Large Telescope of the European Southern Observatory at infrared wavelengths of 1.24 μm (shown in blue), 1.65 μm (green) and 2.16 μm (red). In the center we see four bright and massive young stars, which form the Orion Trapezium, and a large number of less massive stars that formed at the same time, about a million years ago. One also notices the emission by gas and dust illuminated by these stars. © ESO.

Stars result from the collapse of fragments of the initial cloud under their own gravity. During this collapse, the inner parts of these fragments contract faster than the outer parts, which are little affected. The heat resulting from the contraction is initially evacuated by radiation, but there comes a moment when the gas continuously heats up while its density becomes very high. During this time, the protostar continues to grow due to infall of some of the surrounding material. If the mass thus gathered is sufficient (more than 0.08 times the mass of the Sun), the core temperature becomes high enough for nuclear reactions to ignite and the star is born; otherwise, we are dealing with an aborted star, a brown dwarf. The outer parts of the cloud, which has generally some degree of rotation, collapse along the rotation axis to form a circumstellar disc.

However, the central condensation could not lead to a star without loss of a large part of its angular momentum. The solution to this problem appeared unexpectedly in the 1980s, with the discovery of an ejection of matter that systematically accompanies the formation of stars: a symmetrical double jet emerges from the central condensation with a velocity of several hundred kilometers per second and extends over considerable distances. The mechanism of formation of this jet is complex and subtle: it uses the magnetic field of the disc to pump out part of its rotational kinetic energy, which is used to accelerate matter on each side of the disc along its axis of rotation. The ejection of matter from both sides of the disc is a very general phenomenon, which occurs from the very first stages of star formation. The emission of such jets seems to be necessary for the formation of the star-disc system. After a few hundred thousand years, what remains of the cloud where the star was formed disperses, the jets cease and only the central star surrounded by a disc remains, less important than before because part of its matter was ejected or was accreted by the star. It is in this residual disc that planets and comets are formed (figure 2.4).

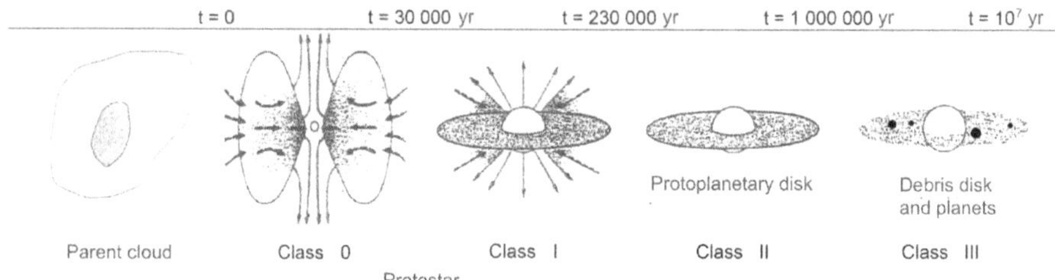

FIG. 2.4 – The stages of the formation of a star and its circumstellar disc. Under the effect of gravity, a part of an interstellar cloud collapses upon itself, faster in its central regions, forming a dense core. When the heat generated by the collapse can no longer be evacuated from this core, the temperature rises there until nuclear reactions are triggered and the star ignites ($t = 0$). While matter continues to be accreted by the star, a bipolar jet evacuates part of the matter with its angular momentum, which slows down the rotation of the star (Class 0 protostar). Then the circumstellar disc is formed, while the jet attenuates (Class I). The jet stops and the accretion decreases; planets begin to be born in the disc that we can now call protoplanetary (Class II). Small dust grains and gas are finally expelled by the wind and the radiation of the star. After a few million years, only a young star and planets remain inside a disc of debris (Class III). Diagram by the authors, from Philippe André.

2.4   The Formation of Terrestrial and Giant Planets

How do the planets appear in the disc surrounding the star in formation? In the 1960s, two different types of models were proposed. In the first one, whose main promoter was the Canadian Alastair Cameron (1925–2005), the disc had a mass close to that of the Sun; the planets formed directly from the gas and dust of this disc by gravitational instability, then a large fraction (about 85%) of the disc matter was swept by the intense wind from the forming Sun, while the rest was accreted by the central star. The less massive planets subsequently lose their gas, retaining mainly the solid or liquid heavy elements.

The second model, that of the Russian Victor Safronov (1917–1999), envisaged a disc of much lower mass, about one hundredth of the mass of the Sun. When the disc cooled down, the heavy elements condensed into dust that stuck together thanks to electrostatic forces to form particles of a few centimeters, then solid bodies of a few kilometers in diameter, the *planetesimals*. These agglomerated to form more massive bodies, the *planetoids*, the seeds of planets. As in the previous model, gas was ejected from the disc, dragging the remaining dust. However, sufficiently massive planetoids could retain the passing gas, giving the giant planets.

Safronov's model is currently favored by astronomers, at least as far as the Solar System is concerned, because many of his predictions are confirmed by observation. However, the gluing mechanisms necessary for the formation of planetesimals are not yet well understood. What followed is better described by numerical simulations: some planetesimals could grow thanks to their gravitational interaction that favors collisions, destruction and accretion. At the end of this scenario dominated by multiple collisions, after a few million years, planetoids were formed. These bodies also collide with each other, merging or fragmenting. The current Solar System is the result of this billiard game.

The nature of the planets formed in this way depended on the nature of the solid matter available to form their nuclei. This was a function of temperature, and thus of the distance to the Sun; indeed, heavy molecules generally condense faster than light molecules (see box 2.1). At the beginning of the history of the Solar System, up to about 3.5 times the current radius of the Earth's orbit in classical models, the temperature due to heating by the proto-Sun was such that water was in the form of vapor, as well as ammonia, methane and other volatile compounds that cannot condense: in this inner region, the planets were therefore mainly formed of refractory elements such as silicates. They were dense, and since the refractory elements formed less than 1% of the mass of the nebula, they were small: this is the case of Mercury, Venus, Earth and Mars, planets essentially made of rocks: they are called the *telluric* or *terrestrial planets*. Farther from the Sun, beyond what is called the *ice line*, water was in the form of ice because the temperature was low; then a little further still other volatile compounds such as ammonia, methane, carbon dioxide and, further still, carbon monoxide were frozen; as water and these compounds are very abundant in the interstellar medium, the planetary nuclei were larger, on the order of 10 Earth masses. Their gravity was such that they could retain hydrogen and helium which made 99% of the mass of the primitive nebula: this is how giant planets like Jupiter and Saturn were formed, with their core of ice and refractory elements surrounded by a region where the hydrogen, which is highly compressed, is in the form of a metallic ocean (recent observations with the Juno space probe show, however, that Jupiter does not have a distinct core), and their gaseous envelope, consisting essentially of hydrogen and helium, which contains most of the mass. The average density of these planets is less than 2 g/cm^3, while that of the telluric planets is much higher.

Box 2.1 The composition of parent matter: interstellar gas and dust. The chemical composition of the matter from which the Solar System was formed is well known, because it is also that of the Sun's photosphere, which was not contaminated by the elements synthesized in the depths of the star; only a few fragile atomic nuclei such as lithium, beryllium and boron have partially disappeared. Concerning the refractory elements, the analysis of a particular class of meteorites, the carbonaceous chondrites, gives information that confirms and even improves the results obtained on the Sun. Table B.2.1 gives the most recent version of this composition, for the elements of most interest here.

TAB. B.2.1 – Abundance of major elements in the Sun's photosphere and in carbonaceous chondrites, from Asplund M., Grevesse N., Sauval A.J., Scott P., *Annual Review of Astronomy & Astrophysics* 47, 481 (2009). Possible discrepancies between the solar and meteorite abundances of carbon, nitrogen and oxygen are due to the fact that these volatile elements only partially condensed into solid form.

Element	Decimal logarithm of solar abundance (in number of atoms, H = 12)	Decimal logarithm of meteorite abundance, normalized to silicon
H	12	
He	10.93 ± 0.01	
C	8.43 ± 0.05	7.39 ± 0.04
N	7.83 ± 0.05	6.26 ± 0.06
O	8.69 ± 0.05	8.40 ± 0.04
Na	6.24 ± 0.04	6.27 ± 0.02
Mg	7.60 ± 0.04	7.53 ± 0.01
Al	6.45 ± 0.03	6.43 ± 0.01
Si	7.51 ± 0.01	7.51 ± 0.01
S	7.12 ± 0.03	7.15 ± 0.02
K	5.03 ± 0.09	5.08 ± 0.02
Ca	6.34 ± 0.04	6.29 ± 0.02
Ti	4.95 ± 0.05	4.91 ± 0.03
Cr	5.64 ± 0.04	5.64 ± 0.01
Mn	5.43 ± 0.04	5.48 ± 0.01
Fe	7.50 ± 0.04	7.45 ± 0.01
Co	4.99 ± 0.07	4.87 ± 0.01
Ni	6.22 ± 0.04	6.20 ± 0.01
Pb	1.75 ± 0.10	2.04 ± 0.03
U		-0.54 ± 0.03

These elements were found in gaseous form and as dust grains in the protosolar nebula. Some, such as rare gases, existed only in gaseous form, although a small amount may have been trapped in the grains. Oxygen, conversely, was both in gaseous form, refractory compounds such as silicates, or volatile compounds such as water, carbon monoxide and dioxide, possibly deposited on the grains as ice or even included in the grains. It is the same for carbon which can be found in gaseous form like methane, or in the form of oxides and many solid compounds. As for nitrogen, it forms many gaseous or solid compounds such as ammonia and the N_2 molecule which can hardly condense on the grains. Carbon and nitrogen are thus strongly deficient in meteorites. This is not the case of oxygen, which is only deficient by an uncertain factor of the order of 2, which means that about half is found in the form of silicates and oxides and the other half in gaseous form: water, carbon monoxide and other rarer oxygenated molecules, with also some molecular oxygen.

Many authors have also determined the abundances of the main elements in the interstellar gas. In the diffuse interstellar medium and even in ordinary interstellar clouds, where there are not many molecules because of the ultraviolet emission of hot stars that partially destroy them. We find that the abundance of gaseous carbon compared to hydrogen is about half of that in the Sun and that of oxygen is about 3/5 of the solar abundance, the rest being found in the grains. On the other hand, nitrogen seems almost entirely gaseous. As for magnesium, silicon and iron, they are very deficient in the gas and are, therefore, almost entirely in the grains. These results are corroborated by the comparison between solar and meteorite abundances presented in table B.2.1, which shows, however, that more of these elements were present in the grains at the time of the formation of the Solar System. The explanation for this is simple: the elements were even more condensed on the grains in the protosolar cloud, which is denser and more opaque than ordinary interstellar clouds. Only the volatile elements carbon, nitrogen and oxygen remained partially in the gaseous state, almost totally for nitrogen.

A particularly interesting case is that of water. This molecule is almost entirely photodissociated in the diffuse interstellar medium and in ordinary interstellar clouds and is almost unobservable there. In denser clouds, it is formed by chemical reactions on the surface of the grains, where it remains as an ice mantle, like many other molecules such as CO. Most of this ice sublimates in stars in formation under the effect of shocks, as indeed observed by the Herschel satellite. However, a part of the H_2O stays included in the grains and is thus still found in carbonaceous chondrites. The rest of the primitive water vapor did not remain in the region where the terrestrial planets were born, as well as all the other gases: during the formation of the Solar System, water was found in solid state only beyond the ice line, where the giant planets and the icy objects, including cometary nuclei, were formed. Meteorites and comets brought more water to the terrestrial planets, especially during the "Late Heavy bombardment": this water formed the Earth's oceans.

2.5   The Migration of Planets

It has long been thought that planets are today at the distance from the Sun where they were born. In fact, this is not the case. As soon as they were formed, their gravitational interaction with what remained of the protoplanetary disc of gas, dust and fragments of various sizes, forced them to move closer or further away from the Sun. This was understood as early as 1980 by Chia-Chiao Lin (1916–2013), John Papaloizou, Peter Goldreich and Scott Tremaine, but the idea looked strange and was slow to take hold; the research on migration within the Solar System was motivated by the discovery of the migration mechanism in exoplanetary systems.

In the case of a massive planet like Jupiter, its gravitational action digs in the protoplanetary disc a circular groove around its orbit, where it is blocked (figure 2.5). It will thus follow the radial motion of the disc matter. In the internal parts of this disc, the planet gets closer to the star as long as the star continues to accrete the residual matter of the disc, and it is the same for small solid bodies and terrestrial planets that are being formed there.

FIG. 2.5 – Numerical simulation of the structure of a protoplanetary disc where a massive planet was formed, after 99 orbital periods of this planet. The density of the disc is represented in false colors, from black to white when increasing. The planet (white dot) creates a denser disturbance (spiral lines) in the protoplanetary disc, a disturbance that slows down the planet and digs an almost empty circular groove in the disc (in black). The planet is then stuck in this groove. © Paris Observatory, LESIA, Philippe Thébault.

Since the beginning of our century, the study of the formation of planets has progressed considerably, notably with the elaboration of the *Nice model* by scientists initially gathered at the Nice observatory. In this model, it is assumed that Jupiter, which would determine most of the subsequent evolution because of its considerable mass, was rapidly forming just beyond the ice line, at a distance from the Sun of 3.5 astronomical units (au, the average Sun–Earth distance); as the amount of matter available there is maximum, it is logical to give birth to Jupiter at this place. The mass of the planet quickly exceeded a hundred terrestrial masses, which led to its migration towards the interior of the Solar System. This migration would not have stopped if another phenomenon had not intervened: Saturn, which was formed at 4.6 au from the Sun, reached in its turn a mass sufficient to trigger also its inward migration, faster than that of Jupiter, until it enters in 3:2 resonance with the latter planet, 3 periods of revolution of Saturn being then equal to 2 periods of Jupiter.

Jupiter was then at the level of the orbit of Mars, at 1.5 au from the Sun. The two planets, which now dug a common groove in the disc, then began a slow outward migration, while remaining blocked in 3:2 resonance, thanks to a mechanism discovered in 2001 by Frédéric Masset and Mark Snellgrove and well reproduced by numerical simulations: if the distant planet (here Saturn) is less massive than the planet close to the star (Jupiter), the transfer of angular momentum with the disc towards the outside of their common groove is less important than the transfer towards the inside, so that the two planets, which remain in resonance, gain angular momentum: they will thus move away from the star. This is what the authors of the model call the "Grand Tack", which causes the two giant planets to go backwards like two racing yachts turning around a beacon. It should be noted that this model is not unique, but it does have the advantage of giving a good account of a certain number of dynamic properties of the small bodies of the Solar System: other scenarios, also resulting from numerical simulation, are currently being studied to try to reproduce as well as possible all the observables that we have at our disposal.

What are the consequences of these migrations on the formation of terrestrial planets? At the time of the Grand Tack, Jupiter had compressed the inner protoplanetary disc to a radius of about 1 au: the telluric planets formed in this disc of planetesimals and planetoids (figure 2.6). Their formation took about a hundred million years, starting from relatively massive planetoids; the last phase was a billiard game between these planetoids, some of which were destroyed while others grew to form the four terrestrial planets. The impact of a large planetoid on the Earth was responsible for the formation of the Moon, between 60 and 100 million years after the formation of Jupiter. The Nice model predicts the low masses of Mercury and Mars compared to the masses of Venus and Earth: Mercury and Mars formed respectively at the inner and outer edges of the planetesimal disc and therefore lacked material; moreover, in the Grand Tack scenario, the growth of Mars was interrupted by the arrival of Jupiter at the time of the "edge turn".

2.6 The Late Heavy Bombardment and Its Consequences

Six hundred thousand years after the formation of Jupiter, there was hardly any gas left around the Sun, many planetesimals had gathered into planetoids and planets that were now fully formed. A large disc of debris remained beyond the orbit of the giant planets. In the Nice model, its mass is assumed equal to 35 times that of the Earth. This disc was formed of the icy planetesimals and planetoids left behind during the formation of the giant planets: ice blocks, dwarf planets and cometary nuclei. Jupiter acted gravitationally on these fragments by ejecting those that were nearby; it thus lost energy and slowly moved closer to the Sun. Saturn and the two other giant planets, Uranus and Neptune, were not able to do this, and gained angular momentum at the expense of the debris, which were mainly beyond their orbit: they, therefore, slowly moved away from the Sun, and the 3:2 resonance between Jupiter and Saturn was broken.

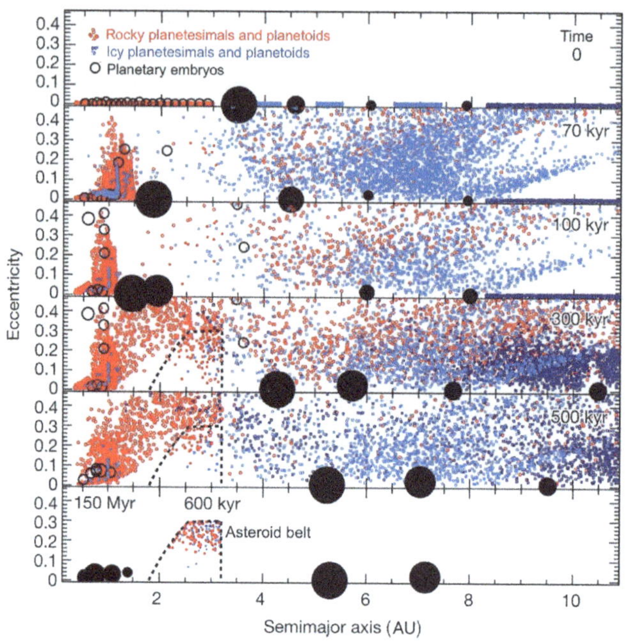

FIG. 2.6 – Possible evolution of the Solar System up to 500 000 years after the formation of Jupiter, based on numerical simulations of the Nice model. Top, the initial state; Jupiter was entirely formed while Saturn, Uranus and Neptune grew larger. 70 000 years later, Jupiter has migrated, pushing in front of it most of the rocky planetesimals and planetoids and some icy ones; most are scattered and acquire important eccentricities. After 100 000 years, Saturn, now completely formed, joined Jupiter in 3:2 resonance and, they started their common outward movement: this is the Great Tack; Jupiter pushed the planetesimal and planetoids to about 1 au from the Sun. After 300 000 years, Jupiter and Saturn have moved apart and are no longer in resonance; the dotted lines delimit what will be the asteroid belt between Mars and Jupiter. At 500 000 years, some of the embryos of rocky planets have grown to gradually form the telluric planets, while others have been, or will be destroyed by collisions. The model successfully reproduces the relative masses of the four telluric planets, in particular the relatively small mass of Mars and Mercury, and the asteroid belt (lower diagrams). The capture of icy planetesimals/planetoids has brought more water to the telluric planets. Adapted from Walsh *et al.*, *Nature* 475, 206 (2011).

A total upheaval took place when Saturn and Jupiter, by moving away from each other, reached the resonance 2:1, Saturn's period of revolution being then twice that of Jupiter: their gravitational effect on the other objects was strongly amplified. According to the Nice model, this event occurred about 880 million years after the formation of Jupiter, or 3.7 billion years ago. The orbits of the giant planets were then deeply modified and took on their current aspect, while those of the terrestrial planets underwent little change. The disc of planetoids was completely dispersed: a part of them ended up as icy bodies outside the Solar System, forming the Kuiper belt and the distant reservoir of comets, the Oort cloud. Another part bombarded the planets and their satellites: this was the *Late Heavy Bombardment.* A major fraction of the craters seen on the Moon, Mercury and the asteroids themselves result from these impacts. Such craters are almost no longer seen on Venus, Earth and Mars because erosion and volcanism have erased them. On the Moon, they dominate the far side, but have been partially covered by volcanic lava on the visible side, where the crust is thinner. It is believed that a large part of the water of the terrestrial planets was brought by these fragments: indeed, these planets were so hot at their formation that the water had then completely evaporated and remained only as inclusions in solids. Since that time, the configuration of the Solar System seems to have undergone few changes.

2.7   The Formation of Planets in Exoplanetary Systems

What do we know about the mechanism of planet formation outside the Solar System? The first discoveries highlighted the importance of the migration phenomenon, which brings giant exoplanets formed far from their star to the immediate vicinity of the star. We will see later (chapter 8) that the giant exoplanets close to their star, the "hot Jupiters", discovered first because they were more easily detectable than the others, are not the majority in the population of exoplanets. Of the 4400 or so known exoplanets, most seem to have a mass between 10 and 30 Earth masses; they belong to two distinct classes: that of the "super-Earths" (the rocky exoplanets) and that of the "Neptunes" (the icy ones). Of course, there is certainly also an observational bias: many rocky exoplanets of mass comparable to that of the Earth still remain to be discovered.

If we compare the structure of the Solar System to that of the exoplanetary systems, we can ask ourselves the question: why did the Solar System not experience a complete migration from Jupiter to the Sun's proximity, as we observe in exoplanetary systems? According to the Nice model and as we have seen, the simultaneous presence of Jupiter and Saturn, with their relative masses and starting positions, prevented Jupiter from approaching in close proximity to the Sun, which would have had the effect of sweeping all the inner Solar System. The probability of such a configuration is undoubtedly very low within the outer systems, which could explain why we have not yet detected, to date, an exoplanetary system comparable to ours.

On the other hand, the good news that exoplanet exploration teaches us is the presence, in very large quantities, of exoplanets with masses less than a dozen Earth masses, many of which could be rocky. We will probably have to wait for one or two decades before we can determine without ambiguity the nature of these objects, thanks to the analysis of their atmosphere, but the current situation leads us to a certain optimism as to the possibility of discovering more habitable exoplanets.

2.8   The Primary Atmospheres of the Terrestrial Planets

Because of their formation scenario and their separation into two classes, rocky planets and giant planets, the atmospheres of the planets are of two quite distinct types. The present atmosphere of the giant planets, composed of a thick envelope of hydrogen and helium, is primary: it was captured at the time of their formation, and the gravity field of the giant planets is sufficient for it to remain stable over the lifetime of the Solar System. Giant planets do not have a surface in the sense that we know: pressure increases inwardly until it reaches several million bars. The case of terrestrial planets is different: their gravity field is insufficient to hold the lightest molecules such as hydrogen and helium. Their present atmosphere, which is secondary, was acquired after the formation of the initial nucleus. It results from two mechanisms: the outgassing of the globe and meteoritic bombardment. There was certainly an atmosphere around the primitive Earth, coming from volatile elements escaping from solids. However, the Late Heavy Bombardment that occurred 800 million years after the formation of the Earth, although it brought little matter and hardly affected the overall chemical composition of the mantle, must have significantly modified this atmosphere.

Can we estimate the composition of planetary atmospheres at the beginning of their history? To do so, we need to be able to define in which chemical form were the most abundant elements after hydrogen and helium, *i.e.* oxygen, carbon and nitrogen. It is possible to do this from thermochemical equilibrium reactions, which allow to predict the relative abundances of the different molecules likely to form. In the case of carbon and nitrogen, these reactions can be written as follows:

$$CH_4 + H_2O \leftrightarrow CO + 3H_2$$

$$2NH_3 \leftrightarrow N_2 + 3H_2$$

These reactions favor the formation of CH_4 and NH_3 at low temperature and high pressure (which correspond to the conditions of giant planets), while they evolve towards the formation of CO and N_2 (and also H_2) under the opposite conditions (which correspond to the conditions of terrestrial planets). It is thus not surprising that the giant planets contain, in addition to hydrogen and helium, the hydrogenated molecules that are methane, ammonia and also some frozen water.

Thermochemical equilibrium therefore predicts that the atmosphere of the terrestrial planets was dominated by CO and N_2, following the escape of hydrogen. Another reaction occurs between carbon monoxide and water vapor:

$$CO + H_2O \leftrightarrow CO_2 + H_2$$

We therefore expect, for the primitive atmosphere of the terrestrial planets, a composition based on carbon dioxide, nitrogen, carbon monoxide and water. We will see that all the information we have suggests, for Venus, the Earth and Mars, a primitive atmosphere rich in carbon dioxide and water vapor, with a minor contribution of molecular nitrogen. We will see further on how these atmospheres, of comparable composition at the origin, evolved according to specific factors towards radically different destinies.

2.9 What Atmospheres for Rocky Exoplanets?

From our experience of the planets of the Solar System, can we imagine what the atmosphere of the exoplanets might be like? Once again, the reactions of thermo-chemical equilibrium give us a first approximation. For exoplanets, we generally know two parameters: the mass and the distance to their host star. Knowing the spectral type and characteristics of this star, we can estimate the radiation it emits, and thus the equilibrium temperature of the exoplanet. This equilibrium temperature can be considered, as a first approximation, as the mean surface temperature of the exoplanet. From these two parameters, it is possible to present a simple classification of exoplanets (figure 2.7).

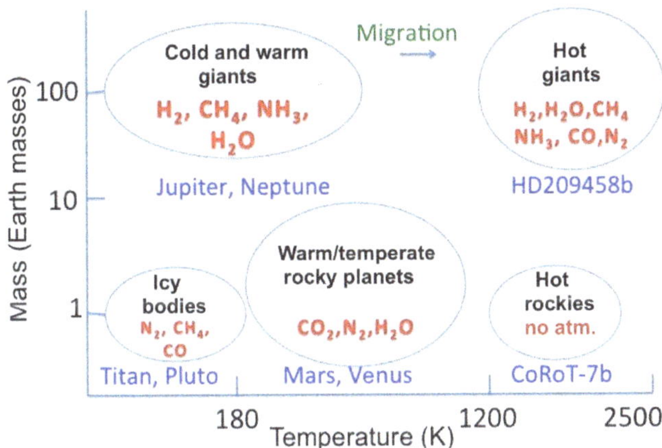

FIG. 2.7 – A simple classification of exoplanets as a function of their temperature and mass. High temperature objects (very close to their host star) have no equivalent in the Solar System. The hot Jupiters are the result of a migration phenomenon. The Earth falls into the category of temperate rocky planets; the composition of its atmosphere has changed due to the presence of oceans, plate tectonics and the appearance of life.

The exoplanets of small mass (less than 10 Earth masses) and low temperature (less than about 200 K, or $-73°C$) are the analogues of the small icy bodies of our Solar System (satellites and trans-Neptunian objects): atmospheres of N_2 (product of the photo-dissociation of NH_3 by the solar UV radiation), CH_4 and CO can be expected. In the same mass range, but with milder temperatures (between 200 and 500 K), we find rocky exoplanets, exo-Earths or super-Earths with atmospheres rich in CO_2, H_2O and N_2 (with perhaps, in case of emergence of life, the appearance of O_2). At higher temperatures, we find small objects closer to their star and devoid of atmosphere: this is the case of Mercury or, as a more extreme example, the exoplanet CoRoT-7b whose day-side surface could be in the form of viscous magma, the temperature exceeding 2000 K.

Let us now move on to the massive planets; we know that they are mainly made up of hydrogen and helium. At low and medium temperatures, we find analogs of our giant planets, where methane and ammonia can be present. At high temperatures, we find the hot Jupiters, unknown in our Solar System but relatively numerous in exoplanetary systems. For the objects closest to their star (at less than 0.05 au for a solar-type star), the thermochemical equilibrium could favor the formation of N_2 and perhaps CO at the expense of CH_4, NH_3 and H_2O.

Of course, this simple classification is only a first draft: thermochemical equilibrium is a simplifying hypothesis, and other mechanisms may intervene, such as photochemistry induced by the irradiation of the host star or dynamic phenomena generating vertical movements. Moreover, let us not forget that the surface temperature of a rocky exoplanet can be different from its equilibrium temperature: this is the case, in particular, of greenhouse planets such as Venus. Finally, planets (and exoplanets) may undergo migrations that may cause their surface temperature, and thus their composition, to change. We will see further on how spectroscopy allows us today, under the most favorable conditions, to determine the atmospheric composition of exoplanets.

Chapter 3

The Exploration of Terrestrial Planets

Because of their proximity to the Earth, the terrestrial planets, visible to the naked eye, have been observed since Antiquity. In the case of Mercury and Venus, their appearance at sunrise or sunset first suggested, in each case, the existence of two distinct objects; then, thanks to repeated observations, each planet was identified as a unique object preceding or following the Sun, and their trajectories were recorded. Their motions were interpreted within the framework of the geocentric model developed by the Greeks and perfected in its most advanced form by Ptolemy in the second century B.C. It was to be an authority for more than fifteen centuries, until the advent of the Copernican era. First proposed by Nicolas Copernicus in his book *De revolutionem Orbium Celestium* published in 1543, the heliocentric model was supported by Galileo's observations and the work of Kepler, which led, in 1687, to the theory of gravitation stated by Isaac Newton in his *Philosophiæ Naturalis Principia Mathematica*.

3.1 The First Astronomical Observations

The observation of the terrestrial planets as physical objects (and not only as point objects in the sky) began in 1610 with Galileo Galilei (1564–1642), when he first turned his new telescope towards the sky. In his *Sidereus Nuncius*, whose fame soon spread beyond the borders, he announced the discovery of the four main satellites of Jupiter, which would later bear his name, as well as the observation of mountains on the Moon. One year later, Galileo discovered the phases of Venus, similar to those of the Moon in orbit around the Earth (figure 3.1). This last discovery was crucial because it showed that Venus is in orbit around the Sun, which is incompatible with the geocentric system of Ptolemy (but not with that of Tycho Brahe). Galileo also noted that the apparent size of Venus varies with its orbital position.

In 1639, the English astronomer Jeremiah Horrock (ca. 1619–1641) described for the first time the transit of Venus in front of the Sun, a rare phenomenon: it occurs in pairs separated by eight years, themselves separated by more than a century.

DOI: 10.1051/978-2-7598-2563-9.c003

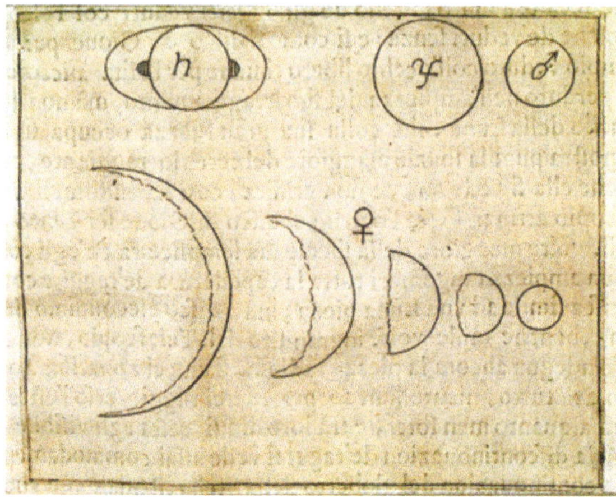

FIG. 3.1 – The phases of Venus as observed by Galileo in 1611 (bottom). They show that Venus must revolve around the Sun, the maximum size being reached when Venus is in front of the Sun (the disc is then almost entirely in shadow) and the minimum size being reached when Venus is beyond the Sun (the disc is then completely illuminated by the Sun). Saturn, Jupiter and Mars are represented on the top. This drawing is taken from Galileo's *Saggiatore* (The Essayist), published in 1623. © Bibliothèque de l'Observatoire de Paris.

Horrock used his observation to determine the diameter of Venus and to obtain, using Kepler's third law, a first, very approximate estimate of the Sun–Earth distance (*i.e.* the astronomical unit, au), about 30% lower than the value accepted today (150 million km), but much higher than what was believed at the time.

The disc of Mars presents structures that were observed by several astronomers, including Christiaan Huygens who, in 1659, drew the first map of the planet; he also deduced its period of rotation, close to 24 h, a value confirmed and more precisely measured by Jean-Dominique Cassini (1624–1712). The latter, appointed in 1671 to the newly created Paris Observatory and an exceptional observer, also studied Venus, but could not determine its rotation period, for lack of structures on the disc that could be used as reference points. On the other hand, he had the idea to use simultaneous observations of the position of Mars with respect to surrounding stars, in Cayenne and Paris, to measure the distance from Mars to Earth and thus deduce the Sun–Earth distance. The result (138 million km) is 10% less than the current value, which is a feat given the limitations of the instrumentation of the time. A more accurate result (within 1%) was obtained in 1769 by observing the transit of Venus from different points on the globe. The previous transit of Venus, in 1761, had allowed the Russian astronomer Mikhael Lomonosov (1711–1765) to infer the probable existence of a dense atmosphere around the planet.

The polar caps of Mars, probably first noted by Huygens in 1672, were systematically studied by several observers, notably Giacomo Maraldi (1665–1729) and William Herschel (1738–1822) who, in 1781, found their variability and precisely measured the rotation period and the inclination of the planet. A little later, the

French astronomer Honoré Flaugergues (1755–1830) discovered the existence of dust storms on the planet. With the improvement of the quality of telescopes, many more observations of Mars followed in the 18th and 19th centuries. In 1877, the American astronomer Asaph Hall (1829–1907) discovered two small satellites around Mars; they were named Phobos and Deimos.

3.2   The Myth of the Martian Canals

In 1877, the Italian astronomer Giovanni Schiaparelli (1835–1910) created a surprise by announcing that the surface of Mars is crossed by straight trails that he called canals ("canali"). Fever seized astronomers such as Percival Lowell (1855–1916), who did not hesitate to attribute these canals to the presence of intelligent beings fighting against drought. However, other observers remained skeptical. This was the case of Edward Maunder (1851–1928), Edward Barnard (1857–1923) and especially Eugène Antoniadi (1870–1944) who, using the large refractor of Meudon, obtained images of unequalled resolution that demonstrated that the "canali" were optical illusions (figure 3.2). However, Lowell and his successors did not disarm and the myth of the Martian canals continued until the advent of the space age. Then, the first images obtained with the Mariner 4, Mariner 9 and Viking probes definitively buried the Martian canals.

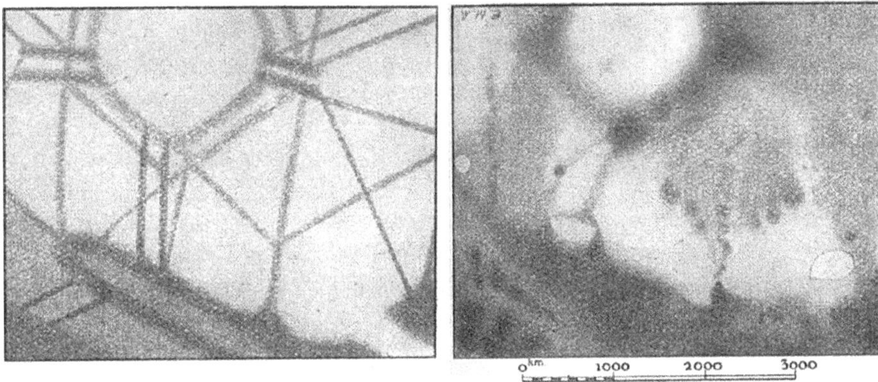

FIG. 3.2 – The "canals" of Mars, observed in 1877 by Schiaparelli (left) and re-observed by Antoniadi with a more powerful instrument in the 1920s (right). Both are drawings. Antoniadi showed that the lines identified as channels by Schiaparelli are only optical illusions. However, some features drawn by Antoniadi are still illusions, when compared to the photographs obtained with the Hubble Space Telescope and space probes. © Bibliothèque de l'Observatoire de Paris.

Schiaparelli did not only observe Mars: at the beginning of the 1880s, he also made drawings of the illuminated part of Mercury and believed he was able to reveal details fixed in time. He concluded that the planet was in synchronous rotation around the Sun, that is, it always turned the same hemisphere towards the Sun, as the Moon does towards the Earth. This conclusion is wrong: we know today, thanks to radar measurements made a century later, that the rotation period of Mercury is 58 days, two thirds of its period of revolution.

3.3   The Physical Nature of Planets

The development of infrared astronomy during the 20th century gave access to another important parameter, the temperature of the planets. In 1924, Edison Pettit (1889–1962) and Seth Nicholson (1891–1963) measured for the first time the temperature of the ground of Mars; it is lower than that of the Earth, especially at the poles. In 1926, they measured on Venus a temperature close to that of the Earth: it is that of the opaque cloud layer. The temperature of Mercury reaches 400°C.

Moreover, with the development of spectroscopy in the first half of the 20th century, it became possible to obtain information on the nature of planetary atmospheres. Water vapor, long sought after on Mars, was detected in the 1950s in very small quantities. Towards the 1960s, it appeared that on Mars and Venus, carbon dioxide is the dominant constituent, while Mercury has no atmosphere. The nature of the polar ice caps on Mars was identified: it is carbon dioxide that condenses alternately at the poles according to the seasonal cycle of the planet. In the 1970s, it was discovered that the clouds of Venus are composed of sulfuric acid, while water vapor is almost absent.

In the middle of the 20th century, the development of radio astronomy also brought its share of surprises. In 1958, the first measurements of Venus at centimeter wavelengths allowed to determine the planet's surface temperature, because radio waves pass through clouds, while the visible and infrared radiations are absorbed by them. The surprise was enormous: the surface temperature of Venus is very high, of the order of 460°C! What could be the cause? At first, astronomers were puzzled; then the idea emerged of an extreme greenhouse effect, due to the carbon dioxide of the very thick and dense atmosphere, an idea developed and confirmed in the following decades. With its torrid atmosphere and clouds of sulfuric acid on the top, Venus could no longer be considered a habitable world, and hopes once again turned to Mars as the exploration of planets was entering the space age.

3.4   The Beginning of the Space Era

The 1960s saw the beginnings of the conquest of space, in a context of fierce competition between the United States and the Soviet Union. The Soviets had the first successes with the launch of the first satellite, Sputnik, in 1958 and the first cosmonaut, Yuri Gagarin (1934–1968), in 1962. The United States replied by launching into the conquest of the Moon, which culminated in the success of the Apollo 11 mission that put the first man on the Moon on July 21, 1969. The lunar samples brought back by the Apollo missions provided valuable information on the age of the

Solar System by defining an absolute dating scale. It should be noted, however, that it was not necessary to send a man to the Moon to obtain these samples, since the Soviets also obtained samples from robotic missions, albeit in much smaller quantities.

The political context was favorable to space exploration of planets. Since the 1960s, beside the conquest of the Moon, Mars and Venus have been the privileged targets of the United States and the Soviet Union, which sent a series of robotic probes to both planets. There were many failures on both sides. On the side of NASA, a first success was obtained in 1962, with the flyby of planet Venus by the probe *Mariner 2*, which confirmed the very high temperature of the surface. In 1964, *Mariner 4* sent the first images of the Martian ground, putting a definitive end to the myth of the Martian canals. The most important following steps in the exploration of Mars were the *Mariner 9* mission in 1972 (figure 3.3), then the two *Viking* probes in 1975, which made a thorough exploration of the surface and atmosphere of Mars. The *Viking* missions, in particular, were a real technical achievement. They consisted of two identical orbiters and two identical descent modules that operated for several years; the resulting database is still a reference today. However, the *Vikings*, despite their immense scientific and technical success, were a great

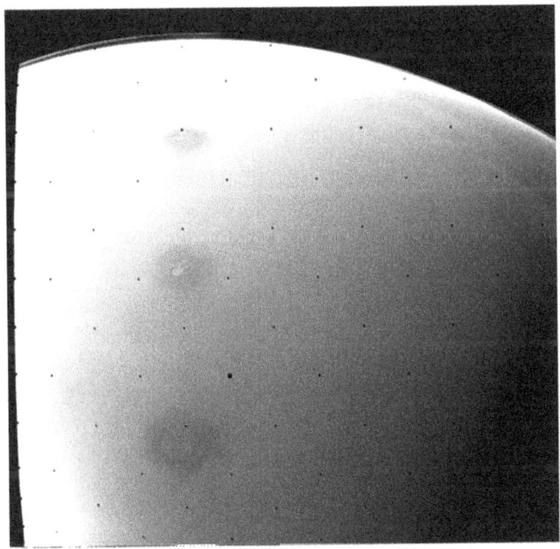

FIG. 3.3 – The first image of the Tharsis volcanoes on Mars, taken by the *Mariner 9* probe in 1972. Launched by NASA on May 30, 1971 and put into orbit around Mars in November 1971, the probe had to wait several months for the end of a global dust storm before being able to send images of the surface to Earth. With the first images sent by Mariner 4 a few years earlier, they contributed to kill definitively the myth of the Martian canals. We see here the three volcanoes of the Tharsis plateau: from bottom to top Arsia Mons, Pavonis Mons and Ascraeus Mons. Their age is larger than 3 billion years and their altitude is between 15 and 18 km. © NASA.

disappointment: they did not find any trace of life, which was their main motivation. The consequences on NASA's planetary exploration program were very important: the Martian program was interrupted for twenty years!

3.5 The Viking Mission: Hopes and Disillusions

Viking was the only space mission to have on board instrumentation that could detect life on a planet other than Earth. Launched in the summer of 1975, the two *Viking* 1 and 2 orbiters entered Martian orbit less than a year later and their two landers descended smoothly. In addition to the classical payload of instruments for the study of the surface and the atmosphere, including the GC–MS (*gas chromatograph and mass spectrometer*) for the detection of organic molecules, they contained three biology experiments dedicated to the search for life (figure 3.4). In the first one, named *Gas Exchange Experiment* (GEE), soil samples were placed in incubation chambers in the presence of water, CO_2 and inert gases to reveal possible traces of metabolism. In the second, called *Labeled Release Experiment* (LRE), a soil sample was placed in the presence of water and a nutrient solution containing glucose with radioactive ^{14}C atoms used as tracers. If microorganisms were present, they would have consumed the nutrient solution and emitted ^{14}C atoms. Finally the third experiment, the *Pyrolytic Release Experiment* (PRE), looked for signs of photosynthesis by putting the Martian soil samples in the presence of light and water in an atmosphere of CO and CO_2, enriched in ^{14}C. Any biomass formation

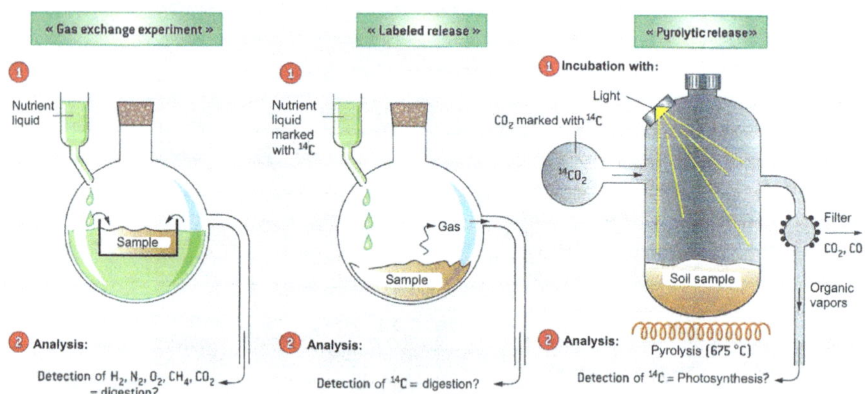

FIG. 3.4 – The instrumentation of the Viking landers for the search for life on Mars. Left: the Gas Exchange Experiment aims at detecting gases produced by Martian microorganisms following the ingestion of a nutrient liquid. Center: the Labeled Release Experiment aims at detecting carbon dioxide, marked with ^{14}C, resulting from the transformation of organic matter. Right: the Pyrolytic Release experiment is looking for the assimilation of CO_2 by the sample following a photosynthesis reaction. Adapted from F. Forget *et al.*, *La planète Mars: Histoire d'un autre monde*, Paris, Belin, 2006.

would then have been detected through the ^{14}C. The experiments were conducted for six years at the *Viking 1* site, and three and a half years at the *Viking 2* site.

The first results were met with great enthusiasm, especially following the unexpected detection of oxygen release in the GEE experiment. However, it turned out that this release remained after sterilization by heating the sample, which finally suggested a chemical origin for this oxygen. The other experiments also provided ambiguous results that gave rise to much debate. The final blow came from the GC–MS chromatograph, which to everyone's surprise failed to detect even the smallest organic molecule at a rate of one part per billion. The most commonly accepted explanation today is that the oxygen produced in the GEE experiment was the result of the chemical oxidation of matter; it thus highlighted the particularly oxidizing nature of the Martian soil. Thus, after long controversies, the scientific community has generally agreed that the Martian soil is devoid of any organic matter, both because of the solar ultraviolet radiation that penetrates the surface and because of its oxidizing nature. One of the potential oxidizing agents is hydrogen peroxide H_2O_2, a product of the dissociation of H_2O by solar ultraviolet radiation. H_2O_2 was discovered much later in the Martian atmosphere, in the early 2000s, with a concentration of less than one part per million.

The *Viking* mission thus showed that the surface of Mars exposed to solar radiation is not favorable for the development of life. However, we cannot completely rule out the possible existence in the past of localized niches, sheltered from sunlight or underground. This is why, when NASA resumed space exploration of Mars in the late 1990s, the focus was on places where liquid water may have been present in the past.

3.6 From Mars to Venus...

Let us go back to the 1970s. The Soviet Union focused on Venus with the *Venera* program. Launched in 1967, the *Venera 4* probe obtained the first measurements of atmospheric composition. In 1970, *Venera 7* was able to land on the surface and measure temperature and pressure – a feat considering the hostile conditions on the surface! The pressure was around 100 bars, and the heat was such that the probes did not survive for long. Later, in 1975, other *Venera* probes sent images of the surface (figure 3.5) and carried out a first radar mapping, revealing large volcanic plains. In 1985, the *Venera* program culminated with the launching of two balloons into the atmosphere of Venus: this was the *Vega* mission (Venera-Halley) made up of two probes that, after having dropped their balloon on Venus, continued to meet Halley's comet in March 1986.

In the meantime, NASA, disappointed by the lack of detection of life on Mars, was again interested in Venus. It launched in 1979 the *Pioneer Venus* mission, equipped with an orbiter and four descent probes that made measurements of atmospheric composition down to an altitude of 12 km. Later, the *Magellan* mission, an orbiter simply equipped with a radar, obtained between 1992 and 1994 a complete cartography of the ground of Venus (figure 3.6). It revealed that the ground is

FIG. 3.5 – The first images of the ground of Venus, taken by the cameras of the *Venera* 9 and 10 landers in 1975. At the bottom of the images, we can see the base of the probe. The ground appears to be covered with basaltic rocks. Other images of the ground of Venus were sent by the following Venera probes between 1980 and 1985. © CCCP.

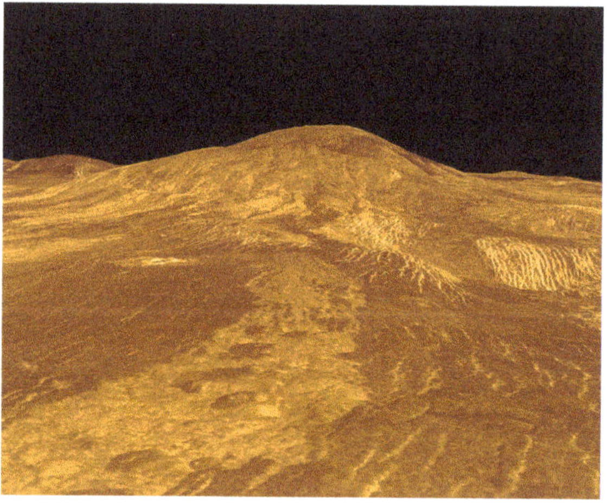

FIG. 3.6 – NASA's *Magellan* mission brought a major advance in our knowledge of the surface of Venus. Thanks to a radar placed in orbit around the planet, capable of probing through the thick clouds of sulfuric acid, the *Magellan* probe mapped almost the entire surface of the planet between 1991 and 1993. This false-color image features Maats Mons, the highest volcano on the planet. © NASA.

entirely covered by relatively recent volcanoes, whose age does not exceed 500 million years. As for the Soviet Union, it also changed its objective after the success of the *Vega* mission: at the end of the 1980s, it turned again to Mars with the

Phobos mission, but with very limited success. Later, Russia launched *Mars-96* and *Phobos-Grunt*, but the political and economic context had changed. These last two attempts were failures, marking the decline of the Soviet space power.

3.7   The Renewal of Martian Exploration

In the late 1990s, NASA was once again looking to Mars, but with a new purpose. If the *Viking* mission did not detect life on the surface of Mars, it cannot be excluded that life existed – or even still exists – in certain privileged sites, protected from the ultraviolet radiation that destroys organic matter. The images from *Viking* have indeed shown the presence of valley networks, suggesting that liquid water may have flowed in abundance in the past. Could Mars' climate have been warmer and wetter in the early history of the planet? NASA's strategy was now built around the slogan "Follow the water!" The idea was to explore sites where liquid water may have stayed, creating a "habitable" environment.

The craze for Mars has also surged thanks to a spectacular announcement made by NASA in 1996: traces of fossilized life would have been discovered in a meteorite from Mars, found in 1984 in Antarctica! Such Martian meteorites had already been identified: they form the SNC group (Shergottite, Nahklite, Chassignite), so called from the name of the site where they fell; there are a little more than a hundred of them, a tiny fraction (about 2 thousandths) of the total population of meteorites. Their Martian origin is established with a very high probability by the similarities in their chemical and isotopic composition with that of the Martian atmosphere made by the *Viking* probes. Following violent impacts, rocks have been ejected from the surface

FIG. 3.7 – The Martian meteorite ALH84001. It was discovered in December 1984 in the Allan Hills in Antarctica (hence its name). Weighing about 2 kg, it has been ejected from Mars about 16 million years ago, and fell to Earth 13 000 years ago. The age of its formation (4.5 billion years) makes ALH84000 the oldest known Martian meteorite. It is famous for the controversy caused by the observation of tiny tubes presenting analogies to nano-bacteria. The explanation generally accepted today by the scientific community is contamination by the terrestrial environment. Wikimedia Commons.

of Mars and travelled through the Solar System; some of them end up caught by the Earth's gravitational field. The studied meteorite, named ALH84001 (figure 3.7) is exceptional by its age. It is one of the oldest meteorites recorded.

The team of David McKay, author of the discovery, announced the presence of traces of fossil life in the meteorite on the basis of several arguments: (1) the presence of carbonate globules and structures close to those of carbonate biogenic terrestrial materials, discovered in the cracks in the rock where water has flowed some 3.6 billion years ago; (2) the presence of complex organic molecules, the *PAHs* (polycyclic aromatic hydrocarbons) that could result from microbial alteration; (3) the existence of particles of magnetite and of iron sulfide that could result from reactions involving biological chemistry, like on Earth; (4) the presence of ovoid or worm-shaped structures (the "nano-bacteria") resembling ancient fossilized microbes, but much smaller than them. None of arguments proposed by the authors gave a definitive proof by themselves, and each clue was the subject of intense controversy. The endogenous origin of the observed biomarkers has been disputed, with some studies concluding for a high probability of contamination of the Antarctic meteorite by surrounding organic material. At present, the scientific community as a whole agrees with this interpretation. The existence of this exceptional meteorite is nevertheless very important for our understanding of Mars' distant past. Formed at the very beginning of the planet's history, in a temperate, humid and reducing environment, ALH84001 tells us that Mars experienced, in its early ages, initial conditions more favorable to life than those of today, since organic molecules were able to stay there without being immediately oxidized and destroyed.

At the end of the 20th century, the stage was thus set for a revival of Martian space exploration. However, the adventure began badly. The *Mars Observer* probe, launched in 1992, lost contact with the Earth as it arrived close to the Martian orbit. This failure prompted NASA to move towards less expensive missions, even if they were riskier: it is the "faster, better, cheaper" concept that led to the launch of two missions at the end of 1996: the *Mars Global Surveyor* orbiter and the *Mars Pathfinder* descent module, equipped with a small robot capable of moving, *Sojourner*. Indeed, to study the environment in detail, it is not sufficient to land at a point on the surface: one must be able to move around to find the most suitable environment. Hence a new generation of missions including orbiter and descent modules, but also surface vehicles. This time, both missions were successful. Two remarkable discoveries have been made by *Mars Global Surveyor*: first, the discovery of a fossil magnetic field in the most ancient terrains of the southern hemisphere, implying the existence of an internal dynamo at the beginning of the planet's history; second, the discovery of a "shoreline", extending at constant altitude for more than a thousand kilometers, which could testify to the presence of a boreal ocean more than three billion years ago (figure 3.8). As for the *Mars Pathfinder* mission, designed mainly for technological purposes, it too fulfilled its mission, and the robot *Sojourner* could move successfully on the surface of Mars.

However, two years later, NASA again experienced a double failure with the *Mars Surveyor 98* mission: the *Mars Climate Orbiter*, launched in December 1998, was lost when put into orbit in September 1999; it was discovered shortly afterwards that the error was due to a confusion between the two different systems of units (the

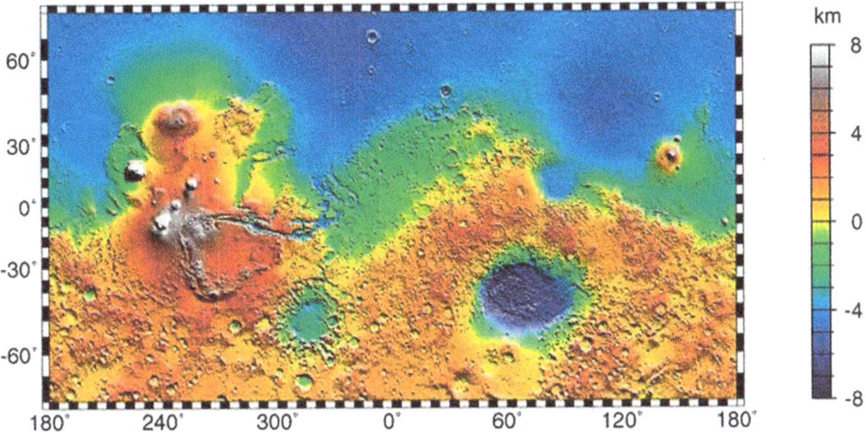

FIG. 3.8 – Altimetry mapping of Mars, performed with the radar altimetry experiment of the *Mars Global Surveyor* mission. It shows that the planet presents a notable asymmetry, the northern hemisphere having an average altitude significantly lower than that of the southern hemisphere; the cause of this asymmetry is not understood. A "dichotomy line", close to the equator, separates the plains of the north from the older reliefs of the south and presents in places a rigorously constant altitude. It is the possible remnant of a shoreline that would have marked the contour of a boreal ocean of liquid water some three billion years ago. © NASA.

American system and the metric system) used by the satellite manufacturers and the NASA navigation officials! The *Mars Polar Lander* probe, launched in January 1999, was also lost in December 1999 when it entered the Martian atmosphere. These failures, in addition to those of the Russian probe *Mars-96* and the Japanese probe *Nozomi* launched in 1998, illustrate that, twenty years after the success of the *Viking* mission, it remains extremely difficult to successfully put a satellite into orbit around Mars, and even more to land on its surface, especially at high latitudes.

Fortunately, with the turn of the millennium, NASA was back to success. First, the *Mars Odyssey* orbiter, launched in 2001, discovered, thanks to its gamma-ray spectrometer, a large quantity of water trapped under the polar ice caps (figure 3.9). Then the year 2003 saw the launch of two identical rovers, *Spirit* and *Opportunity*, which operated for several years, *Spirit* until 2010 and *Opportunity* until 2018. After fifteen years of operation, the *Opportunity* rover had covered more than 45 km until the end of its mission, well beyond the initial mission objectives. The year 2003 also saw the entry of the European Space Agency in the exploration of Mars with the *Mars Express* orbiter, which took over part of the instrumentation of the *Mars-96* mission. In spite of the failure of its *Beagle 2* descent module which crashed on the ground, the *Mars Express* orbiter, still in operation today, is a complete success. Three instruments are particularly innovative: the imaging spectrometer operating in the infrared, which allows to study the mineralogy of the surface, the UV/IR spectrometer observing the Martian atmosphere during solar and stellar occultations, and finally the radar allowing to search for water present under the surface.

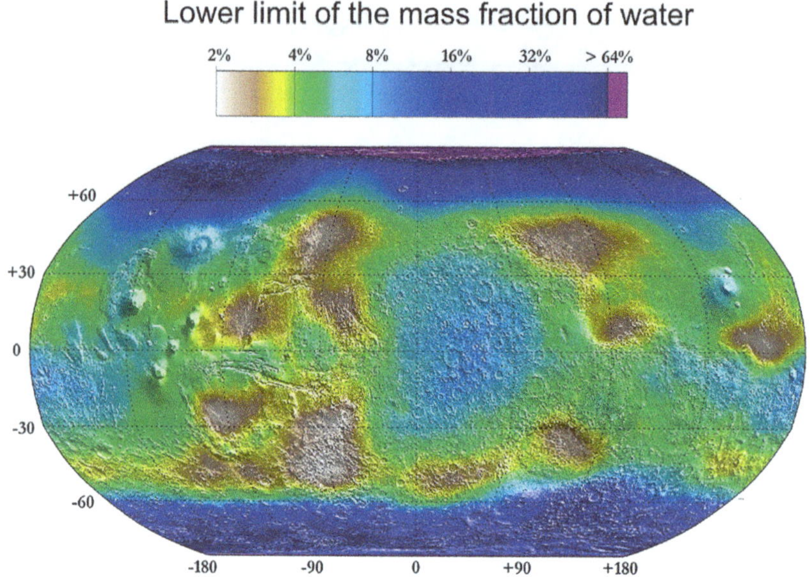

FIG. 3.9 – Mapping of the abundance of water contained under the surface of Mars, obtained by the gamma-ray spectrometer of the *Mars Odyssey* mission. By analysis of the gamma radiation emitted by the surface, the instrument was able to find hydrogen atoms (linked to the presence of the H_2O molecule) in significant quantities, especially under the poles of the planet. Water is trapped under the surface in the form of permafrost or water ice. © NASA.

The complementarity of the measurements of *Mars Express* and the Martian rovers proved to be very important for the study of the Martian mineralogy and the history of water in the early ages of the planet.

NASA's success continued with the *Mars Reconnaissance Orbiter* (MRO), launched in 2005, followed by the *Phoenix* descent module, launched in 2007 and placed near the North Pole, and the *Mars Science Laboratory* launched in 2012, with its *Curiosity* rover, an ambitious mobile laboratory designed to search for traces of past life or, at least, signs of a habitable environment. If it found only very few organic molecules, *Curiosity* discovered stratified sedimentary rocks showing the existence of liquid water and a habitable environment 3–4 billion years ago (figure 3.10). Another remarkable result is the detection of a temporary methane emission, which might have confirmed older and controversial results. In 2013, the *Mars Atmosphere Volatile Evolution* (*MAVEN*) orbiter was launched to measure atmospheric escape; the objective this time is to determine what the primitive atmosphere of the planet may have been in order to better understand the mechanisms of its evolution. The *Curiosity* and *MAVEN* missions are still in operation, as well as *Mars Express* and the *MRO* orbiter that relays and transfers data to Earth.

On the European and Russian side, the continuation of *Mars Express* was prepared jointly. This is the *ExoMars* program, composed of two parts. The first is the *Exomars Trace Gas Orbiter*, designed to search for minor atmospheric gases,

F IG . 3.10 − The Yellowknife Bay site in Gale Crater has been identified by the *Curiosity* robot as a "habitable" environment, *i.e.* one that meets the criteria necessary for the appearance of life: a neutral environment, low salinity (fresh water), the presence of key elements (C, H, N, O, P, S), and the presence of iron and sulfur in different oxidation states. The mineralogical strata bear witness to the presence of a lake some 3–4 billion years ago. © NASA/JPL-Caltech/MSSS.

starting with methane, whose existence is not yet really proven. Launched in March 2016, the spacecraft successfully completed its insertion maneuver in October 2016, but the landing of the *Schiaparelli* descent module ended in failure. The second component is the *ExoMars 2022* rover, currently under development, which should complement the NASA's *Mars 2020 Perseverance* vehicle that has successfully landed on Mars on 21 February 2021. Directly derived from the *Curiosity* rover, the primary objective of *Mars 2020/Perseverance* is the selection and collection of samples from the Martian soil, to be returned to Earth on a subsequent mission called *Mars Sample Return*. This extremely ambitious and costly mission is currently in the definition phase (see chapter 9).

3.8   Return to Venus

Twenty years after the last Soviet mission to Venus – the *Vega* mission with its balloons – and ten years after the American *Magellan* mission, Europe once again turned to Venus. In November 2005, it launched the *Venus Express* mission, which operated in orbit around the planet between 2005 and 2015. Its main objective was to understand the atmospheric circulation of the planet, which is characterized by a longitudinal rotation of the atmosphere faster than that of the surface, with one rotation in 4 days. The *Venus Express* probe also studied the cloud structure and its variability, and looked for clues of a present volcanic activity, in order to better understand the greenhouse effect on the planet. Placed on a very elliptical orbit favoring the close observation of the poles of Venus, the spacecraft was equipped with the classical payload of remote sensing instruments (camera, spectrometers operating in the ultraviolet and infrared, magnetometer), with in addition the possibility for a spectrometer to observe solar and stellar occultations; this technique, very sensitive, allowed the determination of vertical profiles of atmospheric parameters above the cloud layer. Among the main results of *Venus Express,* we can

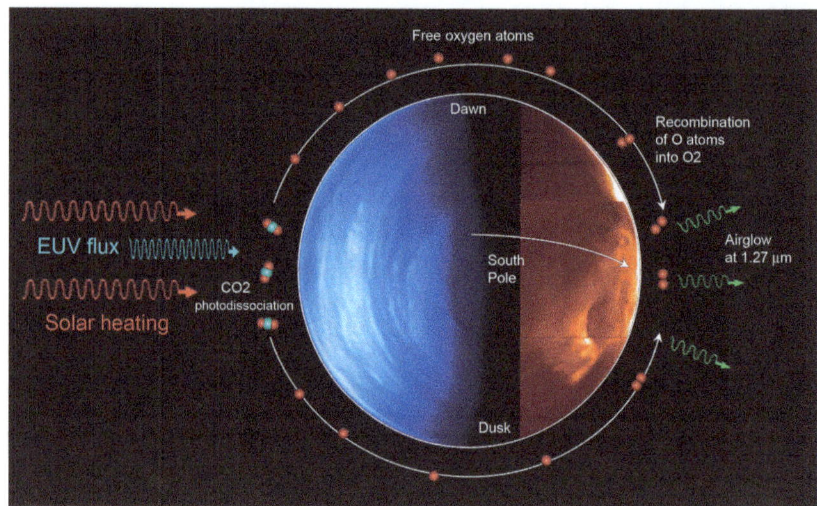

FIG. 3.11 – The *Venus Express* probe, in orbit around the planet Venus, has studied in detail the mechanisms of its atmospheric circulation. In particular, it discovered the existence of a sub-solar – anti-solar circulation, transporting oxygen atoms originating from the photodissociation of carbon dioxide on the day side. These atoms, transported on the night side, recombine in the upper atmosphere to produce excited oxygen O_2, detectable on the night side by its fluorescence radiation at 1.27 μm wavelength. © ESA/VIRTIS/Venus Express.

cite the first observation of the giant vortex of the South Pole, similar to the North Pole one discovered by *Pioneer Venus*, the detailed observation of waves and convection cells present in the cloud layers, the characterization of temperature and velocity fields, the discovery of oxygen and carbon dioxide emissions at high altitude on the night side of the planet (figure 3.11), the probable evidence, in some places, of recent volcanism a few hundred thousand years old at most, and finally the unambiguous discovery of storm lightning observed in the radio domain.

In parallel with the European initiative, the Japanese agency JAXA launched the *Akatsuki* mission in May 2010. Like *Venus Express*, it is an orbiter dedicated to the study of the atmosphere of Venus, this time placed in equatorial orbit to complement European measurements. Unfortunately, in December 2010, *Akatsuki* failed in its insertion maneuver into planetary orbit and went into orbit around the Sun for a period of five years. In December 2015, on the occasion of a new passage near Venus, the Japanese engineers achieved the feat of re-inserting *Akatsuki* into a Venusian orbit, and the probe is still in operation. Thanks to its passages in the immediate vicinity of the equatorial zones, *Akatsuki*'s camera obtains spectacular images of the complexity of the cloud layers and their evolution. It also shows for the first time, in Venus' lower atmosphere, the presence of gravity waves modulated by the surface topography.

3.9 Mars and Venus Today

After fifty years of space exploration, what do we know about our planetary neighbors, Mars and Venus? This is what we will try to summarize below.

With its rotation period and seasonal cycle close to Earth's values, Mars appears to be our closest neighbor in terms of exploration conditions and habitability. Although the temperature is low, the pressure very low and the water vapor almost absent, the diurnal and seasonal cycles are reminiscent of what we know on Earth. The atmospheric evolution of Mars, at the rhythm of the seasonal cycle of two Earth years, is marked by three cycles. The first is that of carbon dioxide, the major atmospheric constituent which condenses alternately to form polar ice caps in winter, inducing variations of 30% in the average atmospheric pressure; the second is that of water which, although not abundant in the gaseous phase (less than a thousandth of the total pressure), also condenses alternately at the poles, creating seasonal ice caps. Finally, the third cycle is that of dust. It is due to the eccentricity (close to 10%) of the Martian orbit, which results in particularly hot summers in the southern hemisphere near the perihelion; the temperature differences between the northern and southern hemispheres generate violent winds, likely to produce dust storms that sometimes cover the entire planet.

The atmospheric composition of Mars is dominated, as we have seen, by carbon dioxide CO_2 and molecular nitrogen N_2. To this are added argon (in the same proportions as N_2), molecular oxygen and other trace constituents, including water vapor H_2O and carbon monoxide CO.

The *Mariner 9* and *Viking* missions have revolutionized our knowledge of Mars. They discovered its enormous volcanoes, the highest in the Solar System, as well as the immense Valles Marineris canyon. They showed that the planet does not have an intrinsic magnetic field; however, the M*ars Global Surveyor* mission discovered the existence of a fossil magnetic field in the oldest terrains, proof of the existence of an internal dynamo at the very beginning of its history. Finally, the space exploration of Mars has also brought to light an enigma that is still unsolved: the history of water on the planet. The first images from space probes have indeed shown the presence of river systems in the oldest terrains, as well as break-up valleys, attesting to catastrophic flooding episodes. The following space missions have brought other clues, all converging in favor of the presence of liquid water in the distant past of the planet; we will come back to this. Liquid water has therefore flowed on the surface of Mars during the history of the planet, but how can we reconcile these facts with the current conditions of the Martian atmosphere that make the presence of liquid water impossible today? We will discuss this in chapter 4.

Although closer to the Earth than Mars, both by its distance and by its dimensions, Venus appears to be radically different from the start. We have seen that the crushing pressure and the torrid temperature that reign on its surface, combined with the presence of a cloud ceiling of sulfuric acid at an altitude of 60 km, make its *in situ* exploration very difficult; moreover, Venus presents other singularities. Unlike the other planets of the Solar System (except for Uranus whose axis of rotation is very close to the plane of the ecliptic), the axis of rotation is oriented at almost 180°

with respect to the ecliptic, which implies that it rotates in the retrograde direction; the reasons for this peculiarity are not clearly established. The planet is rotating very slowly, with a rotation period of 243 days, while its period of revolution around the Sun is 225 days. However, a phenomenon of super-rotation accelerates the winds with a peak at the altitude of the clouds where the period is about 4 days; this phenomenon, undoubtedly related to the very low rotation speed of the globe and the thickness of the atmosphere, is still poorly understood today. Like the Martian atmosphere, the atmosphere of Venus is dominated by CO_2 and N_2, with traces of H_2O and CO. Unlike on Mars, we also find sulfur species: SO_2, OCS and of course the sulfuric acid clouds H_2SO_4. Finally, like Mars, Venus is devoid of magnetic field.

The research conducted today in the context of the exploration of Venus aims at better understanding the mechanisms of its atmospheric circulation, in particular the nature of the vortices that permanently surround the poles. Another question that is being debated is the presence or absence of active volcanism today on the surface of the planet. Indeed, it is extremely difficult to probe the surface of Venus. In the near infrared, there are a few spectroscopic windows through which radiation penetrates to the ground, but scattering by the atmosphere blurs the images, so that no detail at the surface smaller than a hundred kilometers can be seen. Some measurements from *Venus Express* seem to indicate the presence of relatively recent volcanism (less than a few million years). Variations in the abundance of sulfur dioxide, observed at the top of clouds for a few decades, could also be the signature of a present volcanic activity; but this is only a hypothesis. In chapter 4 we will address the question of the evolution of Venus during its history, with a greenhouse effect that radically transformed the primordial Venus to bring it to the present extreme conditions.

3.10   Between Venus and Mars, the Earth

Let us now try to place planet Earth in relation to its neighbors. The history of its exploration has obviously nothing to do with that of Mars and Venus, but it is also marked by stages that have allowed us to discover the nature of its atmosphere and its internal structure.

As for other terrestrial planets, our knowledge of the atmosphere goes back to the first studies of Galileo, followed by measurements of temperature and pressure, made respectively by Evangelista Torricelli (1608–1647) and Blaise Pascal (1623–1662). The atmospheric composition, with its main constituents N_2 and O_2, was discovered at the end of the 18th century, thanks in particular to Antoine Lavoisier (1743–1794). The discoveries multiplied from the 18th to the 19th century, thanks to the sending of manned and sounding balloons, until the discovery of the stratosphere in 1899. In the first half of the 20th century, radio probes made it possible to transmit temperature and pressure measurements from the upper atmosphere, and a monitoring network was set up to study meteorological conditions as far as the stratosphere, to an altitude of about 40 km; this is the level of formation of the ozone layer, the presence of which was revealed in 1913. Finally, the

1960s saw the rise of space exploration, from which our planet was the first to benefit. After rockets that could probe the upper atmosphere for up to 150 km, satellites in terrestrial orbit insure permanent monitoring of meteorological conditions over the entire planet.

The exploration of the Earth's internal structure took longer than that of its atmosphere. If the radius of the Earth has been known since the work of Eratosthenes (ca. 276–194 B.C.), its mass was only determined at the end of the 18th century following the work of Newton, Nevil Maskelyne (1732–1811) and Henry Cavendish (1731–1810). Our knowledge of the Earth's internal structure dates from the 20th century. It was first based on mineralogical and geological studies of the Earth's surface. Mineralogy highlighted the variety of silicates, associating to silicon and oxygen various elements (Al, K, Ca, Fe...) and also water, when they are in hydrated form. Oxides are also abundant, such as iron oxide Fe_2O_3 and calcium carbonate $CaCO_3$, present at the bottom of the oceans. Geological observations have made it possible to study the deformation of folded layers and to probe the Earth's crust (up to 15 km in depth, and up to several hundred km with the study of lava) and the ocean layer (up to 3 km). At the same time, laboratory experiments have allowed the reconstruction of high temperature and pressure conditions (5000°C, 102 GPa or 1 Mbar). Finally, our vision of the interior of our planet has considerably improved during the 20th century, thanks to two major advances, the study of seismic waves and the study of continental drift.

Seismic waves are of two types. Internal waves move with velocities depending on the medium encountered; these are pressure waves that vibrate longitudinally, or shear waves that vibrate transversely. Surface waves move at constant velocity regardless of the nature of the medium they pass through. Since the beginning of the 20th century, networks of seismometers have been set up to study the internal structure of the Earth from signals emitted during earthquakes.

In view of the discussion in the previous chapter on the formation of the terrestrial planets, what can we expect for the composition of the terrestrial globe, as well as that of the neighboring planets? It can be assumed that the elemental chemical composition of telluric planets does not differ from that of the Sun and carbonaceous chondrites as far as refractory elements are concerned. At their formation, the planets were very hot due to the energy brought by the collisions between the solid fragments from which they were formed, and were therefore liquid: it is estimated that the initial temperature of the Earth was about 4700°C. Then, atoms and compounds were sorted by gravity, the heaviest being concentrated in the central regions: this is differentiation. The heaviest elements, iron, cobalt, nickel and the following refractory elements from the Mendeleev chart, accumulated in a central core, while the rest floated above it while cooling. The differentiation concerned all the objects of the inner part of the Solar System large enough for their gravity to be appreciable. The internal structure of the Earth results from this differentiation process, complicated by the gigantic collision that formed the Moon about 50 million years after the birth of the Earth.

The study of the propagation of seismic waves has made it possible to determine the internal structure of the Earth (figure 3.12), and also to know if the medium is solid or liquid. Discontinuities in their propagation speed indicate phase changes.

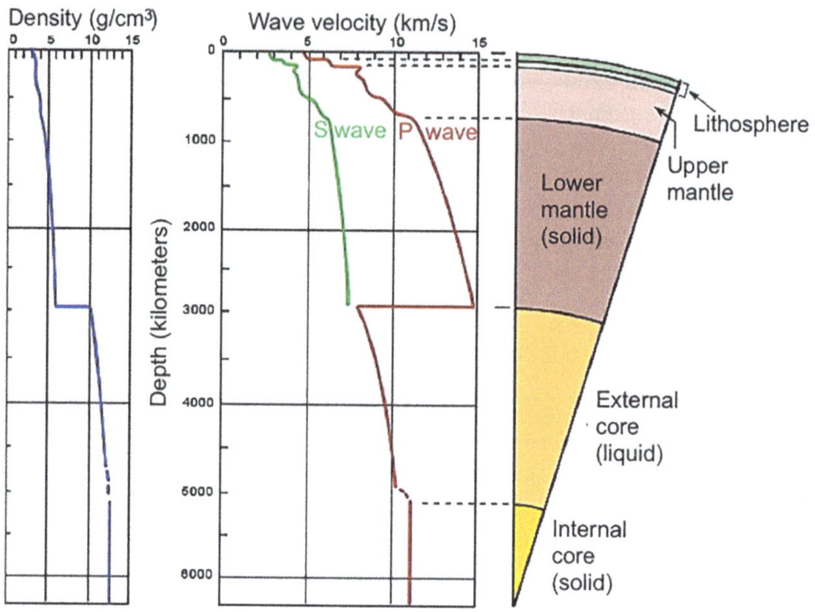

FIG. 3.12 – The internal structure of the Earth. On the left, variation of density with depth. In the middle, velocity of seismic waves as a function of depth. *P* waves are pressure waves similar to sound waves. *S* waves are shear waves in which particles oscillate perpendicular to the direction of wave propagation. On the right, the different layers. The sudden drop in velocity of *P* waves in contact with the mantle and core indicates the transition from the solid state to the liquid state, in which *S* waves cannot exist. The solid part of the Earth's surface (continents and ocean floor), the lithosphere, floats on the upper mantle, whose plasticity allows the continental drifts. Author's drawing, from Bourque.

Thus we can see that the core is solid in its central part due to the very high pressure, up to about 1000 km radius, then becomes liquid up to its limit at 3400 km radius: it is in this liquid part, electrically conductive and animated by convective motions, that the dynamo effect generates the Earth's magnetic field. The mantle that surmounts the core has two layers: an upper mantle 700 km thick surmounting a 2300-km-thick lower mantle. Table B.2.1 shows that the most abundant elements in the mantle are magnesium, silicon and sulfur, with slightly less aluminium and calcium. However, certain very heavy elements such as uranium are also found there because they combined with other elements of the mantle before being able to descend into the core. For similar reasons, some iron also remains in the mantle, and a significant quantity of light elements are found in the nucleus. The mantle elements are mainly in the form of silicates, for example in the upper mantle olivine $(Fe,Mg)_2SiO_4$, with magnesium being dominant. A phase change that produces the differentiation between the upper mantle and the lower mantle, where the olivine decomposes into perovskite $(Fe,Mg)SiO_3$ and magnesiowurstite $(Fe,Mg)O$.

The upper mantle (or asthenosphere) is relatively plastic and exhibits convection that drives the movement of the crust above it, the lithosphere. This is the place where the second great discovery of the 20th century came into play: the continental drift. Its author, Alfred Wegener (1880–1930), based his work on several arguments, in particular the similarity of the coastline on either side of the South Atlantic and of the Indian Ocean (figure 3.13), and the identity of geological formations and rare fossil species present on very distant continents today. However, his plate tectonics hypothesis was (wrongly) refuted in its time, because the mechanism responsible for continental drift – the upper mantle convection – was unknown at that time; it was

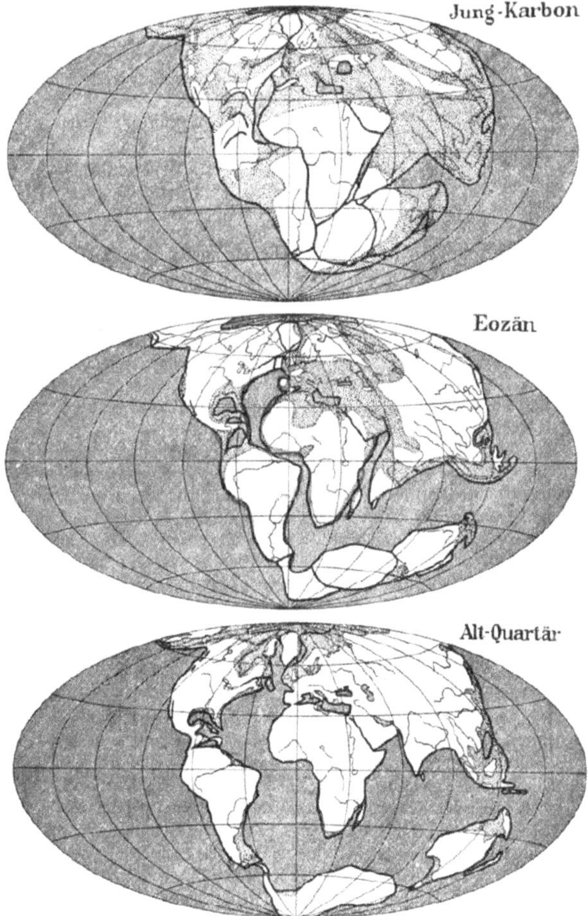

FIG. 3.13 – The continental drift, from Wegener's *The Origin of Continents and Oceans*, 1930. The upper figure is for the Secondary Era (−200 million years), the middle one for the Eocene (−45 million years) and the lower one for the beginning of the Quaternary (−2 million years). Wikimedia Commons.

only accepted in the 1960s with the discovery of the oceanic ridges. Plate tectonics results in the emergence of mantle material – the oceanic lithosphere – at the level of the oceanic ridges, accompanied by the subduction of an equivalent amount of lithosphere under the continental crust. The lithosphere is divided into about fifteen main plates whose boundaries are the seat of earthquakes and volcanic eruptions. Plate tectonics is a singularity peculiar to the Earth: other terrestrial planets are devoid of it, at least at present, and we will see that this phenomenon is of much importance for the habitability of a planet.

3.11 Towards a Comparative Study of the Terrestrial Planets

After several centuries of observation and more than fifty years of space exploration, we have accumulated enough information on Venus, Earth and Mars to be able to attempt a comparative study of these three planets. Some aspects remain poorly known: this is the case of the internal structure of Mars and Venus. The American mission *InSight*, in operation on the surface of Mars since November 2018, should inform us about the internal structure of Mars in the coming years, thanks to the *SEIS* seismograph: small seisms giving S and P waves have been detected and their study is only beginning to give us the requested information. It is nevertheless already possible to draw up a balance sheet of what we know about the three planets, from their interior to their interaction with the interplanetary environment, in order to understand the reasons for their divergent evolutions. This is what we will study in the next chapter.

Chapter 4

Venus, Earth and Mars: A Diverging Evolution

Of the four planets close to the Sun – Mercury, Venus, Earth and Mars – three have stable atmospheres. Mercury, the closest to the Sun and smallest of the telluric planets, has only a transient exosphere, produced by the interaction of solar wind ions with the atoms on the surface. Why is there no neutral atmosphere on Mercury? The reason is its weak gravity field, coupled with the very high surface temperature on the day side. Due to the very slow rotation of the planet and the long daytime duration, the surface temperature there exceeds 400°C; the escape velocity of the molecules is only 4 km/s, insufficient to maintain a neutral atmosphere by gravity. Mercury is therefore not a candidate for habitability in the Solar System.

Let us go back to the three other terrestrial planets, the Earth and its two neighbors, Venus closer to the Sun than the Earth, and Mars further away. Their orbital and physical characteristics are summarized in table 4.1, and table 4.2 shows their atmospheric composition. We have already noted that the three planets present significant differences, both in their physical conditions and in their dynamic properties. First of all, they differ in size and density. While Venus and the Earth have comparable masses, volumes and thus densities, the planet Mars is significantly smaller, with a mass equal to one tenth of the Earth's mass, a radius close to half the Earth's radius and a density significantly lower than that of the Earth. How can this difference be explained? We have seen above that a possible explanation could come from the migration of the giant planets: by moving closer to the Sun, Jupiter would have reached the orbit of Mars before moving outwards under the effect of Saturn's gravity field. The arrival of Jupiter would then have interrupted the growth of Mars by dispersing the planetesimals close to its orbit, before the completion of the planet's accretion phase. This hypothesis has the advantage of accounting for the low mass of Mars, difficult to explain in a classical planetesimal accretion scenario. It is unfortunately difficult to verify, because we cannot go back to the dynamic history of the terrestrial planets in the first billion years; indeed, the solutions of numerical simulations become chaotic... We are thus reduced to hypotheses.

DOI: 10.1051/978-2-7598-2563-9.c004

TAB. 4.1 – Orbital and physical parameters of the planets Venus, Earth and Mars.

Parameter	Venus	Earth	Mars
Semi-major axis	0.72 au	1.00 au	1.52 au
Eccentricity	0.007	0.017	0.093
Inclination over the ecliptic	3.39°	0.00	1.85°
Sidereal revolution period	224.7 days	365.24 days	687.0 days
Mass/terrestrial mass	0.815	1.00	0.107
Equatorial radius/terrestrial radius	0.95	1.00	0.53
Density	5200 kg/m^3	5520 kg/m^3	3930 kg/m^3
Rotation period	243 days	23.93 h	24.6 h
Obliquity	177.33°	23.45°	25.19°
Gravitational acceleration at the equator	8.87 m/s^2	9.81 m/s^2	3.71 m/s^2
Scale height above the surface	15.9 km	8.5 km	10.3 km
Mean pressure at the surface	93.3 bars	1.0 bar	6 mbars
Mean temperature at the surface	735 K (462°C)	288 K (15°C)	215 K (−58°C)
Escape velocity	10.46 km/s	11.2 km/s	5.03 km/s

TAB. 4.2 – Atmospheric composition of the planets Venus, Earth and Mars. Mixing ratios (by volume) are measured in the troposphere (ppm = part per million; ppb = part per billion).

Atmospheric constituent	Venus	Earth	Mars
Carbon dioxide (CO_2)	96.5%	400 ppm	95.7%
Argon (Ar)	70 ppm	0.93%	2.07%
Molecular nitrogen (N_2)	3.5%	78.1%	2.03%
Molecular oxygen (O_2)	–	20.95%	0.173%
Carbon monoxide (CO)	17 ppm	0.01 ppm	0.750%
Water vapor (H_2O)	30 ppm	0.4%	0.03% (variable)
Molecular hydrogen (H_2)	–	550 ppb	15 ppm
Helium (He)	12 ppm	5.24 ppm	10 ppm
Neon (Ne)	7 ppm	18.2 ppm	2.5 ppm
Krypton (Kr)	–	1.14 ppm	300 ppb
Xenon (Xe)	–	90 ppb	80 ppb
Ozone (O_3)		0–70 ppb	40–200 ppb (variable)
Hydrogen peroxide (H_2O_2)	–	–	0–40 ppb (variable)
Nitrogen protoxide (N_2O)	–	300 ppb	–
Nitrogen dioxide (NO_2)	–	20 ppb	–
Methane (CH_4)	–	1.79 ppm	0–40 ppb? (variable)
Sulfur dioxide (SO_2)	150 ppm	–	–
Carbonyl sulfide (OCS)	20 ppm	–	–
Hydrogen sulfide (H_2S)	1–3 ppm	–	
Hydrochloride acid (HCl)	100–600 ppb	–	–
Hydrofluoric acid (HF)	1–5 ppb	–	–
Mean molecular mass	0.0435 kg/mol	0.029 kg/mol	0.0433 kg/mol

4.1 The Astonishing Variety of Terrestrial Planets

Let us start with the appearance of their surface, and what we know about their interior. At 0.7 au from the Sun, the atmosphere of Venus is a furnace, with a ground temperature of about 460°C and a surface pressure of more than 90 bars, or 90 times the average surface atmospheric pressure of the Earth. Its surface, hidden by a thick layer of clouds mainly composed of sulfuric acid, is covered with volcanoes whose existence was revealed to us by the radar of the American spacecraft *Magellan*. The absence of impact craters on the surface indicates that these volcanoes are relatively young, with an average age not exceeding a few hundred million years; in the regions where volcanism is most recent, it is even only a few million years old. Missions to Venus have not detected any sign of tectonics, so the mantle of the planet may not be convective, which would not allow plates of the crust to move. This crust (the lithosphere), rigid and non-deformable, formed of rocks similar to granite and basalt, would be between 10 and 30 km thick. The heat gradually accumulated under the lithosphere could be released periodically in the form of a generalized volcanism producing a global renewal of the surface. Is there active volcanism on Venus today? The question, much debated, has no answer for the moment.

At a greater distance from the Sun, at 1.5 AU, the planet Mars is quite different. Its atmosphere is extremely tenuous, with an average surface pressure of less than one hundredth of a bar. Its temperature is highly variable, oscillating between about −90 and 15°C depending on the seasonal cycle. Because of its small volume, the Martian atmosphere reacts very quickly to fluctuations in solar insolation, whether latitudinal or seasonal; we will come back to this later. The surface of Mars is directly accessible to observation and also has volcanoes: with more than 20 km of altitude, Olympus Mons is the highest peak in the Solar System (figure 4.1). It is a shield volcano of the "hot spot" type, like the ones of the Hawaii islands, and the longevity of its activity testifies to the absence on Mars of plate tectonics similar to that of the Earth. Unlike Venus, Mars also shows the signature of an intense past tectonic activity, with in particular the great Valles Marineris Canyon which extends over a distance greater than a planetary radius. There is no sign of plate tectonics on Mars today, but Valles Marineris could be the signature of its existence in the past. The lithosphere of Mars is very thick, about 50 km on average, and thermally isolates the interior, which loses its heat more slowly than the Earth. This crust is much thicker in the southern hemisphere, where the reliefs are higher, than in the northern hemisphere of the planet, dominated by plains; the reasons for this are poorly understood and are the subject of passionate discussions among specialists.

As mentioned above, the internal structures of Venus and Mars are still poorly known. It is generally believed that the internal structure of Venus is similar to that of the Earth, with a metallic core composed mainly of iron and nickel, topped by a mantle and a crust. The lack of seismological data does not allow to know directly the radius of the core as it was done for the Earth, but a model of the internal structure based on gravimetric data has been proposed: the core would have a radius of 2900 km. Is it liquid or solid? The absence of a magnetic field could suggest that

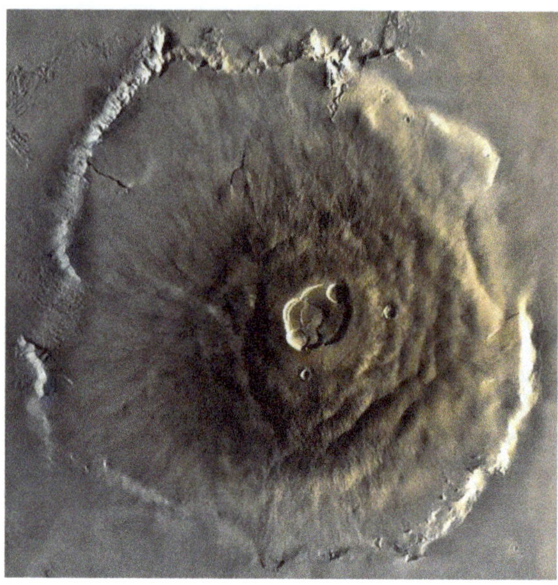

FIG. 4.1 – Olympus Mons, the highest volcano on the planet Mars, observed by the high-resolution camera of the *Mars Express* mission. With an altitude of more than 21 km, it is the highest known volcano in the Solar System. © ESA.

Venus no longer has a liquid core, but it could also be that its rotation is too slow for the dynamo effect to work. This slow rotation also prevents the determination of the moments of inertia of the planet, which could have informed us about its internal structure.

We can think that the internal structure of Mars is the same as that of Earth and Venus. However, it is difficult to say more at the moment. As for Venus, no usable seismic measurements have yet been made on the planet, and we are still eagerly awaiting the results of the SEIS seismometer carried by NASA's *InSight* lander that landed on Mars in November 2018. In the meantime, we only have a standard model based on knowledge of the planet's gravitational field and moment of inertia. The core, whose radius is 1500–1800 km, about half the radius of the planet, seems to contain in addition to iron and nickel a significant amount of lighter elements, especially sulfur, which would lower the melting point and make it liquid. Moreover, Mars possessed a strong magnetic field in the past, which globally disappeared probably at least 4 billion years ago, perhaps due to the disappearance of convective motions in the core that are necessary to produce the dynamo effect. Remnants of this magnetic field remain in some ferromagnetic rocks, in the ancient terrains of the southern hemisphere of Mars. The mantle of Mars is probably quite similar to that of the Earth. Its chemical composition is dominated by iron-rich silicates: the abundance of iron and the presence of many light elements in the core show that the differentiation between the core and the mantle was less extreme than for the Earth.

As for planet Earth, with an average surface pressure of 1 bar and an average temperature of 15°C, it occupies an intermediate position between these two extremes. Seen from space, its surface, partially covered permanently by a layer of water ice clouds, shows alternating continents and oceans of liquid water – a unique property in the Solar System. Similar to its neighbors, it shows intense volcanic activity, but this is still present, as is the plate tectonics that renews the ocean floor in about 200 million years. In addition, the Earth has an intense magnetic field, due to a dynamo effect generated in the liquid outer core of the planet. The interaction of this magnetic field with the solar wind leads to the formation of a magnetosphere, which has the effect of preserving the Earth's atmosphere and protecting the surface from irradiation by the energetic particles of the solar wind, which would be harmful to the appearance and development of life. In conclusion, planet Earth differs from its neighbors in several aspects: the existence of liquid oceans, the magnetic field and plate tectonics.

At the dynamic level, the three planets also present great differences. If their orbits are almost concentric and close to the ecliptic plane (Mars however has an eccentricity close to 10%), the rotation of Venus is retrograde, with almost no seasonal effects (see chapter 3). Unlike Venus, Mars and the Earth have in common an obliquity close to 25° and a rotation period close to 24 h. These similar orbital properties give them similar seasonal behaviors, although exacerbated on Mars because of the very thin atmosphere. On Earth, an ice pack forms around the North Pole in winter (for how long?); on Mars, ice caps of water and carbon dioxide form alternately at the poles according to the seasons.

The interaction with the solar wind also takes different forms for the three planets. The planet Venus is devoid of an intrinsic magnetic field, which can be explained in part by its very slow rotation (insufficient to generate a dynamo effect) and by its internal structure less differentiated than that of the Earth. Venus therefore has no magnetosphere and its ionosphere is in direct contact with the solar wind particles. The Earth, in contrast, has an intense magnetic field that generates a very complex magnetosphere (figure 4.2); polar auroras are a manifestation of this. As for the planet Mars, it does not have a global intrinsic magnetic field. However, at the end of the 1990s, the *Mars Global Surveyor* probe revealed traces of a fossil magnetic field in the oldest rocks, located in the southern hemisphere of the planet. This fossil magnetic field testifies to the past existence of a more intense internal activity, capable of generating a dynamo effect and probably a magnetosphere at the very beginning of the history of the planet. It remains to be understood how and why this internal activity ceased, and what are the consequences for the evolution of the atmosphere during the history of the planet.

Finally, there is a last element that can induce a different behavior between the three planets: the presence or not of satellites. The formation model of the protosolar nebula does not predict the existence of a suite of satellites around the telluric planets. In fact, Mercury and Venus are devoid of them; Mars is surrounded by two very small satellites whose origin is still debated; it could be the capture of small asteroids or the result of an early collision. The case of the Earth–Moon pair is probably due to a catastrophic event, the collision of the proto-Earth with an object the size of Mars. This collision would have ejected in Earth orbit the object in question and part of the material of the young Earth; the fragments would then have agglomerated again to form the Moon. The presence of the Moon around the Earth,

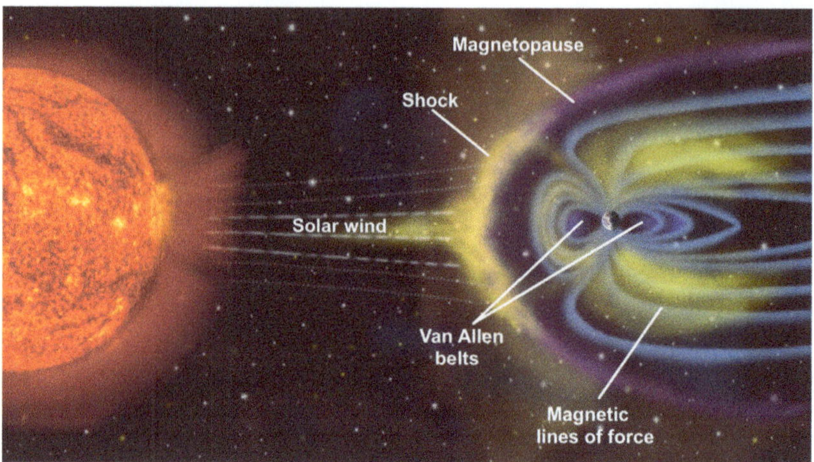

FIG. 4.2 – The Earth's magnetosphere results from the interaction of the solar wind with the Earth's magnetic field. The latter has the structure of a dipole and is generated by the motion of ions and electrons in the outer core of the planet. © NASA.

from the beginning of its history, seems to have had a decisive effect on the evolution of its obliquity. According to recent numerical simulations, the Moon has the effect of stabilizing the obliquity of the Earth, whereas in the case of Mars, calculations show that the obliquity of the planet has in the past experienced very strong periodic fluctuations, up to 60°. Their consequences on the climate have been considerable with, in periods of strong obliquity, the formation of glaciers near the equator. In the case of the Earth, the stabilization of its obliquity over the course of its history may have had decisive consequences on its climatic evolution.

4.2   And Yet... Common Characteristics

Despite their obvious differences, the atmospheres of the terrestrial planets have many things in common. The first one concerns their atmospheric composition. Venus and Mars have atmospheres that are more than 95% carbon dioxide dominated, with a small contribution of molecular nitrogen and, in the case of Mars, argon. The Earth today has a very different atmospheric composition, with 79% molecular nitrogen and 21% molecular oxygen. But this composition does not reflect that of the primitive Earth. We must also take into account the presence in the atmosphere of a very large quantity of water which, in the case of the Earth, condensed during the cooling phase of the planet to form the Earth's oceans. Carbon dioxide, which was also present on the primitive Earth, as on Venus and Mars, in much greater quantities than nitrogen, was then trapped at the bottom of the Earth's oceans in the form of calcium carbonate, *i.e.* limestone. The initial CO_2/N_2 ratio could thus have been comparable on the three terrestrial planets with an atmosphere. The history of water remains to be understood: today it is extremely rare in the atmospheres of Venus and Mars. Very abundant in the primitive Earth, was it also abundant on Venus and Mars at the beginning of their history? We will see further on that this was indeed the case.

4.2.1 The Thermal Structure of the Terrestrial Planets

Like all planetary atmospheres, those of the telluric planets are subject to the hydrostatic law, which expresses the equilibrium between pressure and gravity: the pressure decreases as altitude increases according to an exponential law: $P = P_o e^{-(z/H)}$, P_o being the pressure on the ground, z the altitude and H the scale height, the distance over which the pressure decreases by a factor e (*i.e.* 2.718). The scale height H is equal to RT/mg, where R is the perfect gas constant, T is the absolute temperature, m is the average molecular weight of the atmosphere and g is gravity. The average molecular weight is close to 44 g/mol for Mars and Venus, whose atmosphere is almost entirely composed of carbon dioxide; it is close to 29 g/mol for the Earth, where molecular nitrogen is in majority. At low altitudes, the scale height is equal to 8 km on Earth, 16 km on Venus and 10 km on Mars.

How does temperature vary with altitude? Near the surface, the atmosphere is heated by the surface, which is itself heated by the visible solar radiation to which the atmosphere is transparent. The heat from the ground induces a convection movement that carries energy to the upper layers, to the tropopause, at an altitude of about 12 km on Earth, 50 km on Mars and 60 km on Venus (figure 4.3). At higher

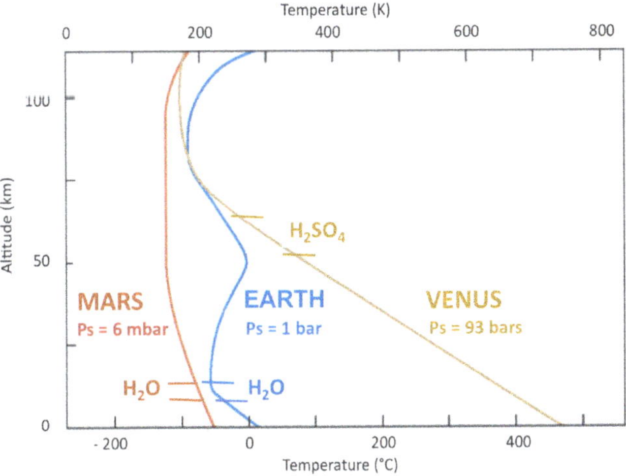

FIG. 4.3 – The thermal structure of the atmospheres of Venus, Mars and Earth as a function of altitude above the surface. All three profiles show a convective zone (the troposphere) extending to the tropopause (at an altitude of 12 km on Earth, 50 km on Mars and 65 km on Venus). Above the tropopause, the Earth's atmosphere shows an increase in temperature linked to the presence of molecular ozone. In the absence of oxygen and thus ozone, Mars and Venus are devoid of stratosphere. Clouds of H2O are present in the tropospheres of Mars and Earth, and clouds of H_2SO_4 are present in that of Venus.

altitudes, the situation evolves differently depending on the planets. On Mars and Venus, a more or less isothermal mesosphere rises to an altitude of about 100 km, at the homopause: this is where atmospheric constituents cease to be uniformly mixed and separate according to their own molecular weight. Various heating mechanisms occur at higher altitudes, including the absorption by atoms and ions of solar UV radiation and the outward propagation of gravity waves. The case of the Earth is different, due to the presence of ozone, a by-product of the photochemistry of molecular oxygen, which absorbs solar UV radiation at an altitude of 40 km, causing a rise in temperature: this is the stratosphere. The ozone layer has the beneficial effect of protecting the Earth's surface from solar UV radiation, which would destroy atmospheric molecules and make life on the surface impossible.

The thermal structure of planetary atmospheres is linked to a cloud structure, due to the condensation of some minor species. One of them is present on the three terrestrial planets, it is water. On Earth, we are familiar with mushroom-shaped cumulus clouds in the troposphere, as well as cirrus trails at higher altitudes. Despite its very low abundance, water also manifests itself on Venus and Mars in the form of condensates. Water ice cirrus clouds appear in the atmosphere of Mars in the colder seasons, and seasonal ice caps form at the poles according to the seasonal cycle. Carbon dioxide cirrus clouds are also sometimes observed at high altitudes or in winter in the polar regions. In the case of Venus, water combines with sulfur dioxide at an altitude of about 60 km to form sulfuric acid, which also condenses to form the thick cloud layer that hides the surface.

4.2.2 Atmospheric Circulation

The first source of energy received by the terrestrial planets is the incident solar flux; however, this flux, averaged over one revolution around the Sun, is maximum at the equator and minimum at the poles; the average temperature thus decreases from the equator towards the poles, inducing a general atmospheric circulation: this is the Hadley circulation found on Venus, the Earth and Mars (figure 4.4).

In the case of Venus, whose obliquity is close to zero, the atmospheric circulation takes the form of stable convection cells: the warm air rises from the equator to the cloud layers located about 60 km away, then moves away towards the high latitudes, cooling down, and falls again at latitudes of 60°, north and south. In the case of Earth and Mars, both of which rotate rapidly, Coriolis acceleration causes a different pattern, with seasonal effects added. On Earth, we observe three cells, the first (Hadley cell) rising from the equator to the tropopause and then descending to the tropics where the trade winds move from west to east, the second one (Ferrel cell) moving in the opposite direction between 30° and 60°, and the third one (polar cell) surrounding the polar regions where the winds again move eastward. On Mars, at the equinox, there is usually a single Hadley cell that rises up to about 50 km and then descends to latitudes of about 50° to the north and south. At the solstice, a single cell is established between the summer hemisphere and the winter hemisphere: the cell originates above the hottest point and descends to the opposite hemisphere.

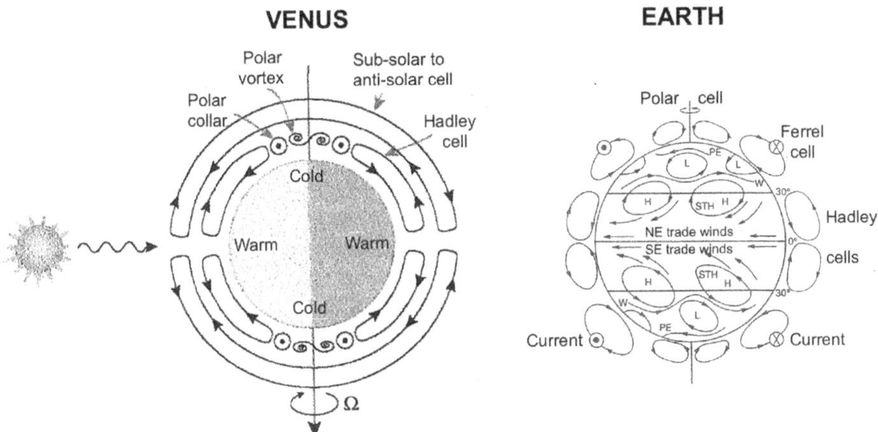

FIG. 4.4 – The atmospheric circulation on Venus (left) and on Earth (right). The Sun is on the left of the figures. In the case of Venus, the rotation speed at the surface is almost zero. In the lower troposphere, only one cell develops in each hemisphere from the equator to high latitudes. A permanent vortex is present near each pole. In the upper troposphere, clouds are subject to a super-rotation with a period of 4 days. Above this level, in the mesosphere, another regime settles in, with a subsolar-antisolar circulation that encompasses the entire planet. On Earth, the Coriolis acceleration generated by the planet's rapid rotation creates a succession of three convective cells in each hemisphere. Adapted from F.W. Taylor, *The Scientific Exploration of Venus*, Cambridge University Press, 2014 and T.E. Graedel and P.J. Crutzen, *Atmospheric Change and Earth System Perspective*, Freeman, 1993.

4.2.3 Internal Structure and Volcanism

We have seen (chapter 2) that the internal structure of Venus and Mars – like that of the Earth and probably also that of the rocky exoplanets – is characterized by a succession of concentric layers ranging from a metallic core in the center to a silicate mantle, containing different silicate compositions.

The internal energy of the terrestrial planets has several origins. On the one hand, it is the energy accumulated during the accretion phase, on the other hand, it comes from differentiation linked to the formation of the iron nucleus; finally, it also comes from the radioactivity of certain radiogenic elements of the nucleus (uranium, thorium, potassium). Heat is transported from the center outward by convective motions that affect both the mantle and the nucleus. The energy released is all the more important as the planet is young: with time, the energy linked to accretion and the radioactive energy gradually vanishes.

The convective motions within the mantle are responsible for the volcanism observed on the three terrestrial planets, as well as the forms of tectonics present on Earth and Mars. On Earth, due to plate tectonics, volcanism can occur either along oceanic faults or in subduction zones. A third type of volcanism originates in "hot spots", probably close to the core-mantle boundary: this is the case of the Hawaiian Islands (figure 4.5). This type of volcanism is also found on Venus and Mars, on an

Fig. 4.5 – An eruption of Kilauea volcano, on the island of Hawaii, photographed on July 16, 2008. Hawaiian volcanoes are characterized by a very fluid lava that spreads rapidly on the flanks of the volcano to form "shield volcanoes". This type of volcano is also found on Venus and Mars. © Michael Poland, U.S. Geological Survey.

even larger scale than on Earth, with in particular the gigantic Mount Olympus on Mars and Mount Maxwell on Venus. These are "shield" type volcanoes, with a gentle slope, whose very fluid basaltic lava can cover vast areas.

4.3   Terrestrial Planets at the Origin

Can we try to identify what the primitive atmospheric composition of the terrestrial planets was? A first approach consists in following the model developed by John Lewis in the 1980s and based on thermochemical equilibrium in the protosolar nebula. This model describes a sequence of condensation as the temperature of the nebula decreases, with the most refractory elements condensing first. We have seen (chapter 2) that this model predicts the formation of reducing atmospheres, based on methane and ammonia, in hydrogen-rich environments (this is the case of giant planets), and oxidizing atmospheres, based on carbon dioxide and molecular nitrogen, in the opposite case (this is the case of terrestrial planets from which hydrogen, too light to be captured, had escaped). We thus find, at first order, the global atmospheric composition of the giant planets and the terrestrial planets.

4.3.1   Secondary Atmospheres

Could the terrestrial planets have acquired their atmosphere directly from the solar nebula? We will see that the answer is no. If the primitive atmosphere of the terrestrial planets was primary, *i.e.* directly accreted by gravity from the protosolar nebula, it should consist of about 63% CO_2, 22% Ne and 10% N_2. However, neon (as well as other rare gases) is almost completely absent on Earth as on neighboring planets. It is therefore clear that even primitive atmospheres are not of primary origin: we speak of secondary atmospheres. Several origins are possible: an outgassing of the globe at the time of the accretion of the planets, an outgassing by volcanism throughout the history of the planets, and finally a contribution by the bombardment of external objects rich in volatile elements, asteroids and comets.

The second hypothesis seems to be invalidated by the measurement of the terrestrial ^{36}Ar/^{40}Ar ratio. Indeed, argon 36 is a primordial isotope, trapped in the planets at low temperature, while argon 40 results from the radioactive decay of potassium 40, with a lifetime of 1.25 billion years. However, the ratio ^{36}Ar/^{40}Ar measured in the gaseous bubbles of volcanic glass shows that argon 36 is more than a hundred times less abundant there than in the Earth's atmosphere. This suggests a rapid outgassing of argon 36 at the beginning of the Earth's history, while argon 40 became progressively enriched in the upper mantle following the disintegration of potassium 40. We will now return to the third hypothesis, that of an external origin of the atmospheres of the terrestrial planets.

4.3.2 Primitive Atmospheres Rich in Water

We have seen that the current atmosphere of Venus and Mars is extremely poor in water vapor. How can we explain this anomaly, when the Earth contains a huge reserve of it within its oceans? We now know the answer: Venus and Mars both had a very water-rich atmosphere at the beginning of their history; this is what we will describe below.

In the case of Venus, the proof came from the measurement of a chemical element, deuterium, an isotope of hydrogen. This atom noted D is a hydrogen atom to which a neutron has been added. It therefore has the same chemical properties as hydrogen, but it is twice as heavy. It can combine with hydrogen and oxygen to form HDO, mono-deuterated water, or heavy water. The D/H ratio, measured in the Earth's oceans, is well known: it is 1.556×10^{-4}. However, in the early 1990s, astronomers were able to measure the HDO/H_2O ratio in the lower atmosphere of Venus, by near-infrared spectroscopic measurement from the Earth (figure 4.6). Confirming an earlier measurement from the *Pioneer Venus* probe, they found that the D/H ratio on Venus was more than 100 times higher than the Earth value! How can this result be explained? The commonly accepted explanation is that water, present in abundance at the beginning of the planet's history, has escaped massively during its history. The photo-dissociation of the H_2O and HDO molecules by solar UV radiation released the H and D atoms that escaped, with the H atom, lighter, leaving more easily.

In the case of Mars, the evidence for an abundant primitive water reservoir is multiple. As in the case of Venus, the HDO/H_2O ratio has been measured many times and leads to an overall enrichment of the order of 5 compared to the terrestrial value. Here again, the reason given is the differential escape which favored hydrogen atoms over deuterium. But there are many other clues that testify to the presence of water, and even liquid water, in the past of the planet (see chapter 3). The first one is the presence of valley networks in the oldest terrains (figure 4.7), identified as early

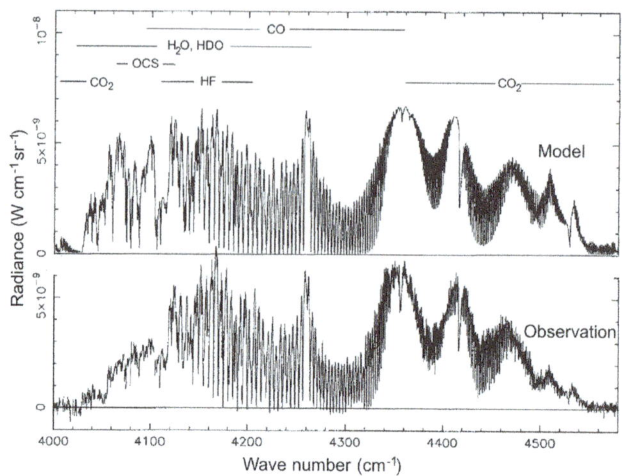

FIG. 4.6 – Measurement of the D/H ratio in the atmosphere of Venus, based on the spectroscopic observation of the H_2O and HDO molecules in the lower atmosphere on the night side of the planet. The spectrum is measured in the near infrared, at wavelengths close to 2.35 μm. The upper part of the figure shows the synthetic spectrum and the lower part the observed spectrum. The molecules responsible for the different absorptions are shown at the top of the figure. The abundances of the molecules taken into account in the synthetic calculation are adjusted in order to obtain the best agreement with the observed spectrum. This one, represented on the figure, corresponds to a ratio HDO/H_2O equal to 120 times the terrestrial value. Adapted from B. Bézard *et al.*, *Nature* 345, 508 (1990).

FIG. 4.7 – An example of a valley network (here Wareggo Valles) on the surface of Mars, observed by the camera of the *Mars Express* probe. These types of reliefs are frequent in the oldest ones and seem to indicate that water flowed in abundance at the beginning of the planet's history. © ESA.

as the 1970s by the *Mariner 9* and *Viking* probes, as well as more recent outflow channels, which have witnessed sudden floods. The second is the discovery of permafrost, especially abundant under the Martian poles; this discovery was made in 2000 by the *Mars Odyssey* probe which detected, using a gamma-ray spectrometer, the presence of hydrogen atoms below the surface at high latitudes (see figure 3.9). The third clue is, in 2005, the discovery of clays in the oldest terrains by the *OMEGA* infrared spectrometer of the *Mars Express* probe; the presence of clays is believed to imply an abundant and durable liquid water flow at the beginning of the planet's history. The *Curiosity* robot, in 2016, discovered sedimentary layers in Yellowknife Bay, part of Gale Crater, indicating the existence of an ancient lake (see figure 3.10). Finally, various clues have suggested the presence of a primitive ocean in the northern part of the planet. The first one is the discovery by the *Mars Global Surveyor* probe in 1998 of a shoreline of constant altitude extending over about 1000 km (see figure 3.8). The second, more recent, is the measurement in this region, by the *Mars Express MARSIS* radar, of the dielectric constant of the ground, implying the presence of water ice under the surface. Finally, the last clue: the *Mars Express* radar detected in 2018 the presence of an aquifer formation under the south polar cap. What could be the importance of Mars' primitive water reservoir? The estimates correspond to a global ocean with an average depth ranging from 100 to 1000 m, depending on the authors and the methods used.

4.3.3 The Paradox of the "Young Sun"

The discoveries brought by the exploration of Mars about the existence of an abundant quantity of water at the origin raised another important question. As early as the 1970s, the discovery of ancient valley networks showed the presence of abundant liquid water, capable of sculpting these reliefs. Moreover, the later formation of outflow channels required the sudden surge of huge quantities of liquid water, perhaps from the rupture of ground water under pressure. The age of ancient valley networks is estimated at -3.8 billion years, and the age of outflow channels at -3.0 to -3.5 billion years.

This is where the paradox of the "Young Sun" comes into play. Models of stellar evolution teach us that the Sun, like all stars of its category staying on the main sequence, sees its luminosity increasing with time. 3.7 billion years ago, its luminosity must have been 70% of its present value. Under these conditions, the equilibrium temperature of Mars (assumed to be in its current orbit) must have been below 205 K ($-68°C$), while the melting temperature of water is 273 K ($0°C$). How could the water then be in liquid form? Many researchers have tackled this problem, but to date no definitive answer has been given. It should also be noted that it also occurs in the case of the Earth in a slightly different form. How could the Earth have escaped a total and definitive glaciation at the time of the young Sun? We will see later (chapter 6) the solutions proposed in the case of the Earth.

Before going any further, let us first ask ourselves the question of the orbit of the terrestrial planets 3.7 billion years ago. Could they have been in orbits closer to the Sun than their current orbits? We have seen (chapter 3) that numerical simulations do not allow us to go back to the earliest ages, before the Late Heavy Bombardment; however, according to the Nice model scenario (see chapter 2), this occurred about 880 million years after the collapse of the solar nebula. After this event, which resulted in the dispersion of a multitude of small bodies, the dynamic models do not predict major orbital changes of the Solar System. However, the traces of river flow on Mars, dating from -3.7 to -3.0 billion years ago, are posterior to the Late Heavy Bombardment. The problem is therefore very real.

What is the most efficient mechanism for heating the surface of a planet beyond its equilibrium temperature? It is the greenhouse effect, which we have already discussed above (chapter 1; see figure 1.2). Let us recall that it occurs in the case of an atmosphere that is transparent in the visible range, but opaque in the infrared: this is the case of the glass walls of which greenhouses are made, hence its name. The surface receives visible solar radiation (which is maximum at the wavelength of $0.5\ \mu$m) which heats it very efficiently. The surface then radiates in the infrared range, but this radiation is absorbed by the lower atmosphere, which in turn heats up, radiates towards the surface and the mechanism is amplified. For the greenhouse effect to take place, the atmosphere must contain gases that absorb in the infrared range; this is the case for water vapor, but also for carbon dioxide and methane. In the case of Mars, we must first estimate what the surface pressure might have been 3.7 billion years ago. Two methods have been used, one based on isotope ratio measurements from the *Viking* probes, the other on the size distribution of the oldest Martian craters. Estimates range from 100 to 900 hPa (0.1–0.9 bar). Unfortunately, the most recent models including CO_2 and H_2O do not succeed in generating a surface temperature higher than 212 K ($-61°$C) due to the greenhouse effect. Other greenhouse gases have also been considered (CH_4, NH_3, SO_2) but without success. To resolve this paradox, saline aqueous solutions were also considered that allow the melting temperature of water to drop to 245 K ($-28°$C); this hypothesis was put forward in an attempt to explain the linear structures observed today on the slopes of the craters and perhaps due to aqueous flows. But even in this hypothesis, the melting temperature remains too high compared to the assumed equilibrium temperature of Mars 3.7 billion years ago.

Finally, another attempt to solve the problem has been proposed: the paleoclimate of Mars would not have been humid and warm, as it was believed for a long time, but cold and dry, occasionally disturbed by "hot" episodes. These episodes could have been caused by giant meteorite impacts, or by volcanic episodes, or both. There is indeed a strong interaction between the atmosphere, the surface and the interior of the planet. Numerical simulations show that the paleoclimate of Mars could oscillate between short "hot" episodes, linked to intense volcanic and/or meteorite activity, and longer "cold" periods, during which the planet would have been inactive.

In the case of Venus, what are the consequences of the "young Sun" hypothesis? Let us assume the planet on its current orbit 3.7 billion years ago. Assuming an albedo (*i.e.* the fraction of solar energy which is reflected) between 0 and 0.5, the equilibrium temperature of Venus on the day side, for a solar radiation equal to 70% of its current value, would be between 278 and 331 K, *i.e.* 5–58°C, a temperature compatible with the liquid phase of water. Venus could thus have known an ocean at the beginning of its history! Unfortunately, we have no trace left, even fossil, because the surface of Venus, as we have seen, is renewed by volcanism in a few hundred million years.

4.4　History of the Terrestrial Planets: A Divergent Evolution

Let us now try to trace the evolution of the atmospheres of our three terrestrial planets. Obviously, there is no common measure between the amounts of information we have in the three cases. In the case of the Earth, we have a mass of archives that we will present later (chapter 6). In the case of Venus, the information is extremely limited because of the recent volcanism of the surface. Between these two extremes, Mars has the advantage of having preserved intact traces of the first ages of its history. On the one hand, its atmosphere is very tenuous, which makes the signatures of meteorite impacts very legible; on the other hand, its surface has not been globally renewed and shows us terrains dating back to 4 billion years ago. Thanks to the multiple probes that have flown over and explored the atmosphere and surface of Mars, researchers have geological, but also mineralogical and chemical clues that allow us to retrace the history of the planet.

4.4.1　Venus: The Ravages of a Runaway Greenhouse Effect

Is it possible to estimate the amount of water present in the primitive atmosphere of Venus? The measurement of the D/H ratio is ambiguous. On the one hand, the enrichment of the D/H ratio by a factor greater than 100 with respect to the terrestrial value provides us with a lower limit of the initial water reservoir: if only hydrogen atoms escaped, the equivalent of the initial reservoir corresponds to about 5 m of global average water thickness. In reality, the massive escape of hydrogen was probably accompanied by an escape of deuterium atoms, in a proportion that is difficult to quantify. Moreover, the D/H value measured today could refer to a reservoir resulting from meteorite impacts that have marked the history of the planet, and would then have no link with the primordial reservoir. There is therefore an extreme uncertainty on this parameter. According to the authors, the primordial water reservoir of Venus could be from 0.02 to 5 times the current terrestrial value. A quantity of water equal to that of the Earth today and assumed entirely in gaseous form would correspond on Venus to a partial pressure of water vapor higher than 100 bars.

As early as the early 1970s, studies showed that the very high surface temperature of Venus could be explained by a runaway greenhouse effect during its history. In 1972, the first calculations of the atmospheric evolution of Venus, the Earth and Mars were carried out based on simple hypotheses. These calculations assumed planets initially devoid of atmosphere and a progressive degassing by volcanism in a ratio $H_2O/CO_2 = 4$, without taking into account the gradual warming of the Sun. In the case of Venus, the calculations indicate that water is permanently in vapor phase. As the pressure rises, the greenhouse effect gets out of control, as it is fed by both water vapor and carbon dioxide, and the surface temperature rises rapidly. The case of the Earth is different because its surface temperature allows water to condense into liquid form, which regulates the greenhouse effect; we will see below how. As for Mars, in this model, the surface temperature is low enough at the beginning so that the water passes from the vapor phase to the ice phase as the pressure rises. We know that, in the case of Mars, this simplistic model is inaccurate since it is now proven that liquid water has flowed on Mars in the past; the error, in this case, comes from the assumption of a continuous outgassing of the atmosphere over time; we will come back to this later. What is interesting in the calculation is that it gives a good account of the evolution of the atmosphere of Venus under the action of a runaway greenhouse effect. Subsequent improvements made by many authors, including the consideration of the rise over time of the solar flux, have not significantly modified the conclusions of this calculation in the case of Venus and the Earth, even if it is now accepted that water may have existed in liquid form on Venus in the past.

Let us note, however, that the starting hypothesis of the above calculation, considering planets *a priori* devoid of atmosphere and subject to a progressive outgassing, is probably incorrect. It is thought today that the atmospheres of terrestrial planets – secondary atmospheres as we have seen – come partly from gases included in the globe during the accretion phase and partly from external inputs. The latter may have occurred throughout the history of the planets, but particularly during the Late Heavy Bombardment about 3.7 billion years ago.

Venus was probably once rich in water. How and why did all this water disappear? The responsible mechanism, as we have seen, is the photodissociation of water vapor in the upper atmosphere of Venus, followed by the escape of hydrogen (and deuterium) atoms. The fate of the oxygen atoms is not clear; despite their high mass, they escaped with the hydrogen at the very beginning of the planet's history, when the UV radiation emitted by the Sun was more intense than today. They may have been trapped in the surface and contributed to the oxidation of the crust. Finally, the near-total disappearance of water from Venus may be related to the absence of plate tectonics, whereas the Earth, with its mass and radius very close, has them. Indeed, water plays an important role in mantle deformations by reducing the viscosity of olivine and promoting the setting of individual rigid plates.

4.4.2   Mars: A Planet on the Verge of Geological Extinction

The planet Mars stands out from its two neighbors by a much smaller mass, a tenth of that of Earth. Its gravity field is therefore weaker, which implies less meteorite bombardment. Its internal energy, linked to the quantity of radioactive elements contained in the globe, is also lower, resulting in a more limited volcanic and tectonic activity, which, above all, has decreased rapidly over time.

To retrace the history of Mars, planetary scientists have defined, as on Earth, large geological periods. The Noachian extended from the origins to the Late Heavy

Bombardment, which is dated to about −3.7 billion years ago. The ancient valley networks, the formation of the Tharsis Shield, the Hellas Basin, date from this period. Then came the Hesperian, between −3.7 and −3.0 billion years ago, which saw the appearance of the outflow channels with perhaps, just after the Late Heavy Bombardment, the appearance of a boreal ocean. During the Amazon, less than 3 billion years ago, volcanic activity gradually vanished.

Measurements in infrared spectroscopy by the recent space probes, *Mars Express* and *Mars Reconnaissance Orbiter,* have made it possible to associate these periods with a particular mineralogy. They have indeed detected the presence of clays in the oldest terrains of the southern hemisphere, while sulfates have been identified in the more chaotic terrains or near the break-up valleys. We mentioned above that these results implied, in some places, the presence of abundant liquid water during the Noachian, and also the existence of violent and transient flows, possibly related to volcanic activity, during the Hesperian.

Finally, the discovery by *Mars Global Surveyor,* in 1998, of a fossil magnetic field in ancient terrains (see chapter 3), brought an additional piece to the puzzle of the history of Mars. During the first billion years, the accretion energy of the globe and the energy from radioactive elements were sufficient to liquefy part of the interior and generate a dynamo effect and thus a magnetic field. The young magnetosphere of Mars then prevented the escape of its primitive atmosphere, probably largely outgassed from the globe. As the reservoir of radioactive elements of the globe was exhausted, the dynamo stopped and the magnetosphere disappeared, making possible the escape of a large part of the atmosphere. It is also possible that the atmosphere was partially swept away as a result of a severe meteorite impact. The greenhouse effect fueled by carbon dioxide also decreased, causing a drop in temperature and the storage of water in the form of permafrost and polar ice caps.

If today the Martian water ice condenses at the poles, as it does on Earth, it is because the obliquities of the two planets are very close to each other. However, this has not always been the case: indeed, recent numerical simulations have shown that the obliquity of the planet Mars varies according to a cycle of about 120 000 years, the amplitude of these variations being moreover modulated with a period of 2.4 million years: the obliquity can go from 0° to 60°, and the effect of its variations on the climate is very important. While ice remains confined to the poles when the obliquity is low, the situation is reversed for obliquities greater than 40°. Since the poles receive then on average more solar energy than low latitudes, glaciers migrate towards the equator. Remnants of these glaciers have been identified on space probe images at the locations predicted by the global climate models, thus giving them a decisive validation.

4.4.3 *The Earth, Ideally Located in Relation to the Sun*

The paradox of the "young Sun", mentioned above in relation to Mars and Venus, also applies to the Earth: by what mechanism did our planet escape the "global snowball" scenario? If the early Sun was indeed fainter than today by 70%, an insufficient equilibrium temperature would have led to a partial glaciation of the continents, whose albedo would have increased, resulting in a further decrease of the surface temperature. As we shall see later (chapter 6), it is probably the volcanism and the outgassing of carbon dioxide that has created a greenhouse effect sufficient to break out of this vicious circle. Once the oceans are in place, a reaction comes into play that is of crucial importance: in the presence of silicates, carbon dioxide reacts with liquid water molecules to form calcium carbonate $CaCO_3$. This mechanism has allowed the sequestration of carbon dioxide at the level of the ocean crust and, thanks to plate tectonics, the sequestration of carbon in the mantle. This is how the Earth escaped the runaway greenhouse effect that struck Venus...

We now have to understand why and how the terrestrial atmosphere evolved over its history, from an initial composition dominated by carbon dioxide to its present composition, rich in molecular nitrogen and oxygen. It is the apparition of life on Earth that is responsible for this transformation. To understand it, we need to wonder what life is, and to analyze how it appeared on our planet; this is the subject of the next chapter.

Chapter 5

The Appearance of Life

5.1 What is Life?

How can we define life? We only know about life on Earth, and our concepts are necessarily biased by this unique example. Terrestrial life could be only a special case, just as the Solar System seems to be a rather special case among planetary systems. And it is quite risky to define a phenomenon on the basis of a single case. However, if we want to try to detect and characterize forms of life on other planets, we must try to define what we mean by "life" or "the living".

On Earth, the elementary unit of living beings is the cell: a compartment within which biochemical reactions (metabolism) take place, allowing these beings to draw energy from their environment, to subsist, to grow, and to reproduce autonomously. During reproduction, mutations may appear that allow evolution and adaptation to changes in the environment.

From these characteristics, scientists have tried to deduce a definition of life. Let's say right away that there is no consensus on a single definition. The main characteristics of life that are most often cited are those on Earth: differentiation relative to the surrounding environment, capacity to use the energy of the environment, reproduction and evolution, and possibility of adaptation. The question of energy is central: terrestrial organisms derive their energy from complex molecules, which they construct from the elements present in their environment. However, most of the chemical reactions involved are not specific to living organisms. Reproduction is also important: this is how we can identify an organism, because its shape is constant over successive generations. None of these criteria alone is sufficient to define life, because there are always systems that we consider as non-living that fulfill them. Crystals can grow and multiply, but no one will call them alive. Clays, which are varieties of silicates organized in sheets, adapt to their environment, for example, by storing water in the space between the sheets when the environment is humid, and evolve by changing their structure according to the availability of minerals. The case of viruses is particularly thorny, since they are unable to

DOI: 10.1051/978-2-7598-2563-9.c005

reproduce autonomously without infecting a cell. But if we consider that viruses are not alive, what about some bacteria that have also lost this ability to reproduce autonomously and behave like parasites? Finally, the capacity for evolution is not a directly operational criterion, since it presupposes being able to observe entities over a time period compatible with this evolution.

This discussion on the definition of life is not just an academic exercise: when we try to detect signs of life on a planet other than Earth, we are unlikely to come across multicellular organisms that move, feed and reproduce under the eyes of our detectors. It is much more probable that we will be dealing with very subtle traces of potential microorganisms, and we will be directly confronted with the fact that, at the level of its elementary constituents, it is difficult to differentiate the living from the non-living. This is one of the reasons why it has been so difficult to interpret the Martian biology experiments of the Viking missions, as explained in chapter 3.

That it is so difficult to define life suggests that the boundary between living and non-living entities is not completely clear-cut, and that perhaps the transition between the two has been gradual. This is what we will look at now.

5.2   From Spontaneous Generation to Primordial Soup

Scientists currently believe that life results from the action of the laws of physics and chemistry on suitable materials, if the environment is appropriate. Although we do not have any experimental verification of this belief, it is at the basis of all research in exobiology, even if it is not always explicit. But this idea is relatively recent, and the question of the origin of life has not always been posed in these terms: until the 19th century, the commonly accepted theory was that of "spontaneous generation".

Then, the general idea was that in addition to the usual reproductive mechanisms, by which a living being is born by reproduction of organisms of the same species, fully formed beings can appear spontaneously, usually from mud or rotting matter. This is a very old conception: for example, the pre-Socratic philosopher Anaximander of Miletus, in the 6th century B.C., thought that fish were born from the action of the Sun on a humid environment. Aristotle, in his *History of Animals*, observed that insects and worms often appear in the morning dew and that mice, frogs and fish seem to be born from the mud of rivers. He, therefore, deduced that creatures can appear by spontaneous generation, without the intervention of two parents. Spontaneous generation remained the commonly accepted theory for the origin of life until the 17th century. This is evidenced by the affirmation of Lepidus, a character in Shakespeare's play *Antony and Cleopatra*, published in 1623: "Your serpent of Egypt is bred now of your mud by the operation of your sun; so is your crocodile". To which Antony answers: "They are so".

In the 17th century, this hypothesis became a subject of debate and experimentation. The physician Jean-Baptiste van Helmont (1577–1644) observed that by laying out a dirty garment with flour for 21 days, the flour turned into an adult mouse. However, further experiments by Francesco Redi (1628–1698) showed that rotting meat did not produce maggots if protected with gauze. The ensuing controversy gave rise to many variations of these experiments, with different culture media and different ways of sealing or unsealing the containers. With the microscopic observation of microorganisms, bacteria and spermatozoa by the Dutch scientist Antoni van Leeuwenhoek (1632–1723), one began to suspect that the matter was more complicated.

In the 19th century, no one believed anymore in the spontaneous generation of adult animals, but the generation of microorganisms was still a controversial issue. The French Academy of Sciences thus offered a prize of 2500 francs to "anyone who, through well-conducted experiments, will shed new light on the question of so-called spontaneous generation". The prize was awarded to Louis Pasteur (1822–1895) in 1862. The "well conducted" experiment consisted of placing a cultured broth in a glass balloon with a long swan's beak, which trapped outside dust without preventing air circulation. When heated to high temperature, the broth remained sterile, but as soon as it was put in communication with the outside, microorganisms developed in it. As Pasteur concluded: "Spontaneous generation is a chimera: every time we believed in it, we were the plaything of a mistake".

But if life does not appear by spontaneous generation of already formed organisms, how did living beings appear? How can we move from the question of spontaneous generation to that of the origin of life? Apart from divine intervention, there are only two possible answers. One is that the first living beings were brought to Earth from other planets by comets or meteorites, and that others came at various times: this is the theory of *panspermia*, which had its hour of glory. Defended in particular by the chemist and Nobel laureate Svante Arrhenius (1859–1927), it is based on the fact that certain bacteria appear to be resilient enough to survive a long journey into space. However, it only displaces the problem, because life had to appear elsewhere, for example on Mars, from which we receive from time to time a fragment expelled by the fall of an asteroid or a large meteorite. This did not seem to be a problem to Arrhenius who thought that like matter, life was eternal, and that very small microorganisms could leave the surface of planets under the effect of auroral electric fields. The other possible hypothesis is that life appeared on Earth as a result of a long process that gradually passed from inert matter to microorganisms through simpler entities, through mechanisms to be specified. In a way, it is also spontaneous generation, but of much simpler organisms than today's living beings and on very long time scales. This idea had to wait until the end of the 19th century, when the very great antiquity of the Earth became obvious, thanks in particular to the dating work carried out by Henri Becquerel (1852–1908) based on the radioactivity of the oldest rocks.

The idea of an origin of life derives quite naturally from Charles Darwin's (1809–1882) theory of evolution. Although he remained rather cautious on this subject in his publications, he wrote on 1 February 1871 to a friend, the naturalist Joseph Hooker: "It is often said that all the conditions for the first production of a living organism are now present, which could ever have been present. But if (and oh what a big if) we could conceive in some warm little pond with all sorts of ammonia and phosphoric salts, light, heat, electricity etc. present, that a protein compound was chemically formed, ready to undergo still more complex changes, at the present day such matter [would] be instantly devoured, or absorbed, which would not have been the case before living creatures were formed". Of course, Darwin would not have dared to present this idea publicly.

In the twentieth century, the idea of Darwin's "little warm pond" spread out. In the 1920s, the Soviet biochemist Alexander Oparin (1894–1980) and the British biochemist John Burdon Sanderson Haldane (1892–1964) independently proposed theories suggesting that life had an origin in the Earth's distant past, an origin that could only be explained by the laws of physics and chemistry. The two scientists started from the observation that the synthesis of organic compounds from inorganic materials is possible, whether in the laboratory or in the natural environment (the first synthesis of urea dates back to 1828, that of amino acids to the 1850s), and proposed that the appearance of the first living being is the result of a long chemical evolution.

This proposition is now known as the Oparine–Haldane hypothesis. To summarize it in the words of Haldane: "When ultraviolet light [from the Sun] acts on a mixture of water, carbon dioxide, and ammonia, a vast variety of organic molecules are made, including sugars and apparently some of the materials from which proteins are built up... In the present world, such substances, if left about, decay – that is to say, they are destroyed by microorganisms. But before the origin of life, they must have accumulated till the primitive oceans reached the consistency of a hot dilute soup... The first precursors of life found food available in considerable quantities, and had no competitors in the struggle for existence... The first living or half-living things were probably large molecules synthesized under the influence of the Sun's radiation, and only capable of reproduction in the favorable environment in which they originated". In the primordial soup, the organic molecules would thus evolve with increasing complexity to become half-living and then living entities, within protocells called *coacervats*. Haldane suggested that viruses represented an intermediate step in the transition from the prebiotic soup to the first heterotrophic cells. Life may have remained, he wrote, "in the virus stage for many millions of years before a suitable assemblage of elementary units was brought together in the first cell".

The ideas of Oparine and Haldane differ on several points and, in particular, on the form taken by carbon in the primitive atmosphere: methane for Oparine, who stressed the importance of hydrocarbons as basic building blocks, carbon dioxide for Haldane. We will see that this is an important point. The details of the processes that lead from organic molecules to the complex molecules of life were left very unclear, which is not surprising: on the one hand, the problem is still largely with us today, and on the other hand, the structure of DNA was only discovered in 1953.

5.3 The First Experiments in Prebiotic Chemistry

The hypothesis of Oparine–Haldane laid the foundations of prebiotic chemistry and opened up the possibility of experimenting on the origin of life: is it possible, in conditions close to those of the primitive Earth, to reproduce Darwin's "little warm pond" in the laboratory, and to synthesize some of the elementary building blocks of life? After a number of not very conclusive trials, a first answer came in 1953 with the famous experiment by Stanley Miller (1930–2007) and Harold Urey (1893–1981). Let us recall its principle (figure 5.1). A balloon contains a mixture of gases representing the primitive atmosphere according to the ideas of the time, a mixture of methane, ammonia and hydrogen, therefore a reducing atmosphere. In another balloon, supposed to represent a warm primitive ocean, water is boiled; the water vapor circulates in the apparatus, mixes with the gases of the atmosphere in the balloon, where electrodes produce electrical discharges (these simulate thunderstorm lightning). The water vapor condenses (this is the rain), carrying with it the compounds that may be synthesized if they are not volatile, returns to the ocean, and the process starts again.

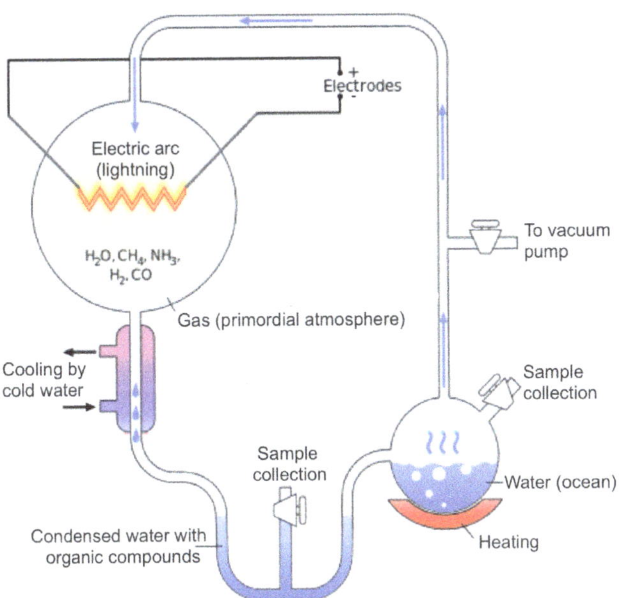

FIG. 5.1 – Diagram of Stanley Miller's experiment. The balloon containing hot water simulates the primitive ocean (right). The balloon representing the primitive atmosphere, supposedly reducing, is on the left. Gases, water vapor, methane, and ammonia are subjected to electrical discharges (lightning). The synthesized products condense in water and return to the ocean. Adapted from Wikimedia Commons.

After a few days, Stanley Miller noticed a change in the color of the condensed liquid, and after a week of continuous operation, 10–20% of the initial carbon had formed organic products, including five amino acids, but no sugars, lipids, or nucleic acids. In 2007, after Miller's death, the contents of sealed containers containing the product of these experiments were re-analyzed by one of his students, Geoffrey Bada, who found that at least 25 different amino acids had in fact been synthesized.

Miller performed variants of his experiment, including one in which a steam jet was directed at the electric discharge, which was supposed to simulate the conditions of a volcanic eruption. It produced 22 amino acids, 5 amines and many hydroxyl radical molecules, more than his other experiments. In another 1961 experiment by Juan Oró (1923–2004), hydrogen cyanide HCN, a molecule very abundant in the Universe that is fundamental for prebiotic chemistry, in aqueous solution with ammonia, produced amino acids and even adenine, one of the four bases of DNA; the other three bases were later synthesized in similar experiments.

Miller and Urey's seminal experiment shows us that an atmosphere of methane, hydrogen and ammonia, subjected to adequate energy inputs, produces a mixture of simple organic molecules. However, it does not solve the question of the origin of life. In particular, it is now known that the Earth's primitive atmosphere was not reductive when life appeared, but was composed mainly of carbon dioxide, nitrogen and water vapor (see chapter 6). The experiment was replicated under these conditions, and the production of organic compounds was then much lower. More fundamentally, it tells us nothing about how living organisms are produced from this mixture, or even how the macromolecules that are essential for life on Earth are produced.

5.4   The Building Blocks of Terrestrial Life

Let us therefore review some of the essential characteristics of life on Earth, which any scenario for the appearance of life must take into account. The first is the importance of water. Living tissues are composed mostly of water, in the order of 70% for most of them. Most biochemical reactions take place in the aqueous phase. If water plays such a fundamental role, it is because it is an excellent polar solvent: in a water molecule, although globally neutral, the hydrogen atoms carry a positive charge while the oxygen atom carries a negative charge, hence a dipolar electrical moment. This allows water molecules to not only interact with each other but also with any other polar molecule through low energy bonds called hydrogen bonds. The presence of liquid water implies a temperature between 0 and about 100°C at the usual pressures.

Biochemistry is essentially carbon chemistry in the presence of water, but other chemical elements are absolutely necessary. They are few in number: with hydrogen (H), oxygen (O), carbon (C), nitrogen (N), sulfur (S) and phosphorus (P) (often summarized under the acronym CHNOPS), we have practically all the chemistry of living things. This chemistry involves a large number of macromolecules or polymers, ranging from a few dozen to millions of monomers. The most important families of these macromolecules are proteins, sugars, nucleic acids and lipids.

Proteins and peptides are polymers composed of amino acids (amino acids are molecules containing both an acid group, the carboxyl –COOH, and an amine group, $-NH_2$). Only 20 different amino acids are involved in living terrestrial matter, whether animal or vegetal. The combination between amino acids is done by condensation, *i.e.* elimination of a molecule and formation of a bond, the peptide bond (figure 5.2). When the molecule so obtained contains up to 10 amino acids, we speak of a peptide, a polypeptide if they are between 11 and 20 acids, and a protein above 21 acids. For example, the egg white protein corresponds to a polypeptide chain of 129 monomers belonging to these 20 amino acids. Some of these chains can spiral, a structure favored by the presence of hydrogen bonds between the =NH group of one peptide unit and the =CO group of another one.

Proteins play an important role in living cells as hormones or catalysts (they are called enzymes in this case), tissue supports (collagens), etc.

Sugars are molecules that contain many hydroxyl (–OH) groups, and come in the form of rings with 5 or 6 carbon atoms when dissolved in water. They can form long chains of polysaccharides, such as cellulose. These compounds play an important role for energy storage or as a structural element of cells.

Nucleic acids such as DNA, which contains the genetic information, have a central role in protein synthesis. They are very long chains of millions or even hundreds of millions of nucleotides, themselves constructed from a 5-carbon sugar molecule, a phosphate group (PO_4^{3-}), and a nitrogen compound called the nitrogen base or nucleic base (figure 5.3). Nucleotides have structural analogies with

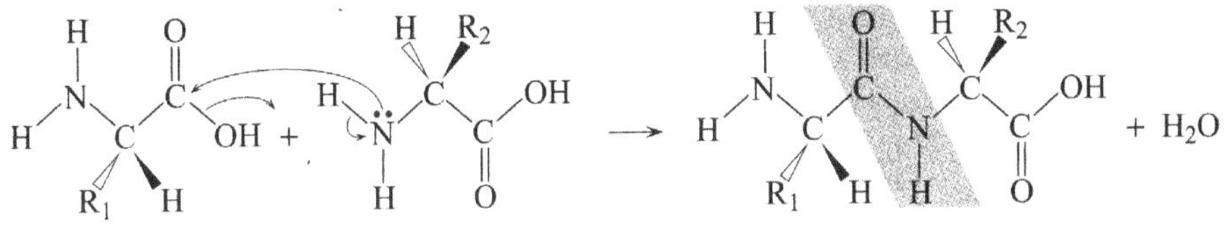

FIG. 5.2 – Formation of a peptide bond by combination of two amino acids. R1 and R2 are characteristic radicals of each amino acid (*e.g.* –H for glycine, –CH$_2$OH for serine). The arrows on the left indicate the detailed mechanism of the reaction. The amide group –CO–NH– in gray is the peptide bond that joins the two sections together. From Luft (2014).

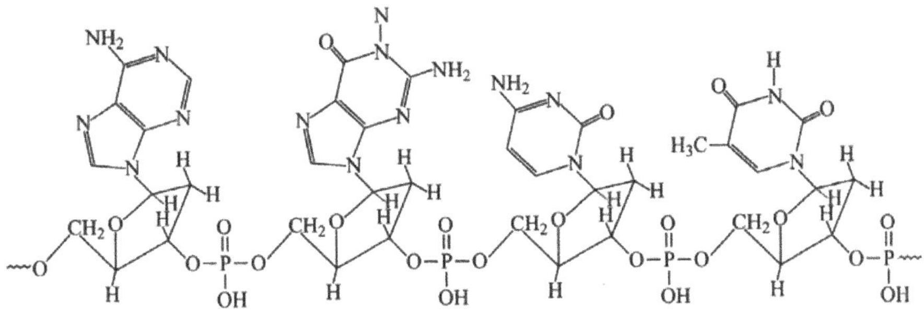

adenine (A)   guanine (G)   cytosine (C)   thymine (T)

FIG. 5.3 – An example of a basic structural fragment of DNA, containing the four nucleotides linked together by phosphate groups. Carbon atoms are not shown (with exceptions) to lighten the presentation. From Luft (2014).

peptides, and like peptides, they are capable of assembling into long helical chains. The most famous nucleic acid is the deoxyribonucleic acid, DNA (figure 5.4), which contains four nucleotides, each with the same sugar but with a different nitrogen base: adenine (A), guanine (G), cytosine (C) and thymine (T). The order in which these nucleotides are arranged defines the genetic code. The RNA (ribonucleic acid) is very similar to DNA, but with a different sugar, and the nitrogen base uracil (U) instead of thymine. RNA has the ability to provide both genetic information and catalysis, something that DNA cannot do; on the other hand, this molecule is made up of a single strand, containing short sequences of nucleotides, whereas DNA, with its double helix, containing millions of nucleotides, is capable of self-replicating.

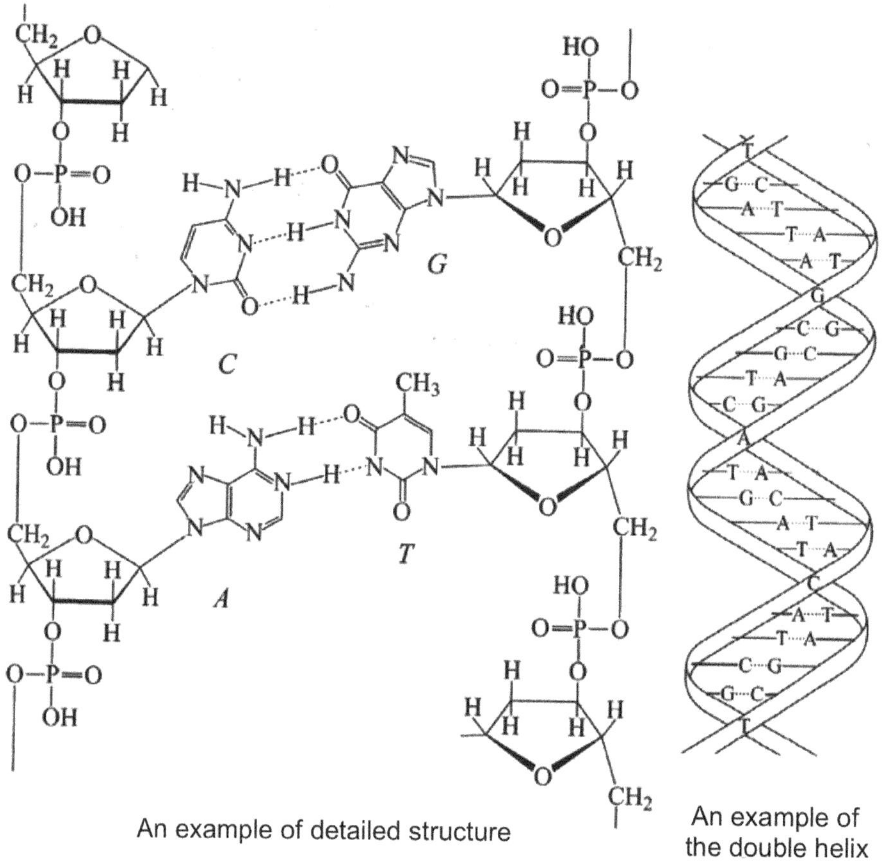

An example of detailed structure

An example of
the double helix

FIG. 5.4 – A fragment of the DNA double helix. Carbon atoms are not shown to lighten the presentation, with some exceptions. The four nucleotides of figure 5.3 (A, G, C, and T) are shown. The two helices are linked by N–C bonds on the right and hydrogen bonds on the left (dotted lines). On the right, schematic view showing a possible arrangement of the nucleotides. From Luft (2014).

Finally, lipids are macromolecules composed of long carbon chains. They are important constituents of cell membranes. They are very frequently phosphoglycerides (figure 5.5). Lipids are amphiphilic, *i.e.* they tend to hold on by one end (head), which has a dipole moment, to water molecules that also have a dipole moment. The other end is hydrophobic. Globally, they are not very soluble in water: in the presence of liquid water, lipids stick together perpendicularly to the surface so that the hydrophilic heads are in contact with water, the hydrophobic tails remaining in the air. A monomolecular film is thus formed on the surface of water.

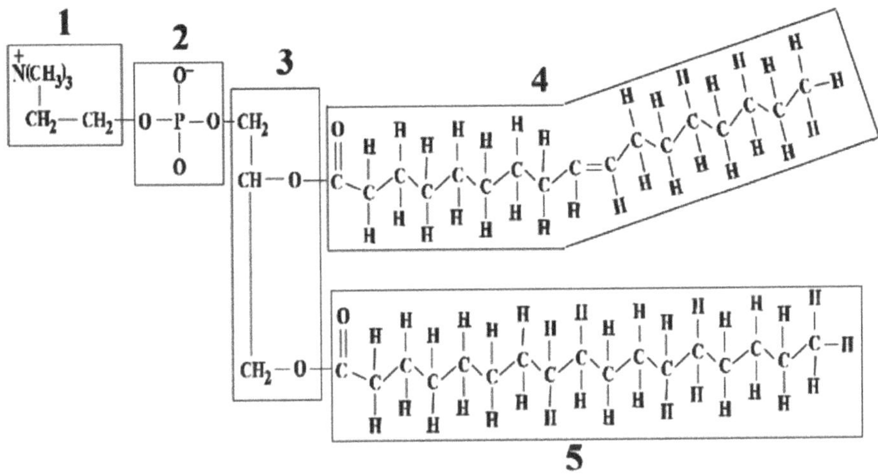

FIG. 5.5 – Formula of a typical phosphoglyceride. Group 1 is choline, group 2 is phosphate and group 3 is glycerol; together they form the hydrophilic head, which has a dipole moment thanks to the + and − charges. Tails 4 and 5 are fatty acids, respectively, unsaturated and saturated, which are hydrophobic. Wikimedia commons, Danntzikg.

Any scenario for the origin of life on Earth must therefore explain not only the formation of prebiotic molecules but also that of macromolecules of living organisms and cells. It is also necessary to understand how the first living beings drew energy from their environment, and finally to propose a mechanism to ensure their reproduction and evolution.

5.5   The Origin of Prebiotic Molecules

The first step is to understand the origin of the simplest prebiotic molecules, and in particular that of the monomers that serve as building blocks for macromolecules. The Miller–Urey experiment has shown that this step can take place at a very fast time scale (of the order of a week), but under conditions that are not those of the primitive Earth as a whole. One possibility is that the synthesis of the prebiotic molecules took place in regions of the primitive Earth where the local conditions would have been comparable to those of Miler and Urey's experiment: for example, underwater hydrothermal vents, where volcanism produces methane CH_4, ammonia NH_3 and hydrogen in abundance. These hydrothermal vents explored since the 1980s are home to a luxuriant life, despite the absence of light and oxygen, high temperatures and a generally acidic environment (figure 5.6), and in particular many chemo-synthetic microorganisms. Given the "extreme" environmental conditions, these organisms are said to be extremophiles: hyper-thermophiles are those that accept a temperature of 100°C or more, acidophiles live at very acidic pHs, of the order of 3... Their presence testifies to the great adaptability of life, but we shall see that it may not be unrelated to the origin of life.

FIG. 5.6 – A white smoker emitting fumes rich in barium, calcium, silica and carbon dioxide in the "Champagne" source of the Marianas Pit. Wikimedia Commons.

Prebiotic molecules could also have come from space. Indeed, such molecules exist in the interstellar medium where more than 200 molecules comprising up to 11 atoms have already been discovered, including, for example, a simple sugar, glycol aldehyde, and also in comets and asteroids which are made of interstellar matter with little chemical modification. A very serious possibility is therefore the contribution of prebiotic molecules by asteroids and comets. A recent study based on the chemical composition of comet 67P/Churyumov-Gerasimenko (figure 5.7), observed by the European probe *Rosetta*, suggests that comets may have provided a considerable amount of prebiotic material, the mass of which could be equivalent to the total biomass present today on our globe (about 10^{12} tons). No less than 58 different molecules, mainly organic, were detected in this comet, including an amino acid, glycine, and products that are particularly chemically active and essential for prebiotic chemistry, such as hydrocyanic acid HCN and formaldehyde H_2CO. This comet also contains a large amount of refractory organic matter, perhaps mainly

FIG. 5.7 – Comet 67P/Churyumov-Gerasimenko, photographed by the camera of the space probe *Rosetta* during its exploration of the comet in August 2014. Many complex organic molecules were identified in this comet, in particular an amino acid, glycine. © ESA.

polyoxymethylene, a macromolecule that results from the polymerization of formaldehyde, and a similar amount of mineral matter, mainly silicates.

Some meteorites, the carbonaceous chondrites, contain a material apparently similar to cometary matter, but annealed, and could have participated in seeding the Earth. Take for example, the Murchison meteorite that fell to Earth in September 1969. Observers who were the first to arrive on site reported the smell of solvent emanating from the meteorite, a first indication of the presence of organic matter, confirmed by the analyses carried out in the laboratories set up for the Apollo Lunar Samples: an incredible variety of organic molecules, including amino acids. In fact, the composition of Murchison proved to be very similar to that of the result of the Miller and Urey experiment! In addition to more than 70 different amino acids, including 5 of the 20 biological amino acids, the Murchison meteorite contains simple sugars. However, it does not contain polysaccharides; it contains nucleic bases, but no nucleic acids.

It should be noted that while nearly 80 different amino acids have been found in meteorites, only 8 of the 20 that are necessary for life on Earth are present. Let us also note that no peptide has been found in the interstellar medium, comets and meteorites; no trace therefore of the formation of organic macromolecules.

5.6   The Rise of Complexity from Prebiotic Molecules

Whether the prebiotic molecules arrived from space or were formed *in-situ* in the atmosphere and then put into solution, or near hot hydrothermal springs, there are still many steps before being able to build the molecules of living organisms. One of the difficulties is to concentrate the basic elements sufficiently so that they can combine. From this point of view, the ocean is undoubtedly too vast and diluted, and scientists rather imagine that these steps took place in puddles or pools of water: this is Darwin's "little hot pond". The process can be facilitated by the fact that these ponds are periodically drained; in fact, the synthesis of macromolecules through the formation of peptide bonds involves the loss of a water molecule, which makes the products relatively unstable in aqueous solution. We can also think of intertidal zones (also called foreshores), those parts of the shoreline that the tide uncovers and then covers: this idea is particularly interesting when we think that at the time when life appeared on Earth, the Earth and the Moon were closer than now and, therefore, the tides were much stronger. Other concentration mechanisms have been considered, such as the freezing of aqueous solutions (because it is water that freezes first), or the surface of clays, which acts as a trap for organic matter.

This hypothesis on the role of clays was defended by the English crystallographer John Desmond Bernal (1901–1971) in the 1950s and developed in particular by the chemist and biologist Alexander Graham Cairns-Smith (1931–2016). Indeed, clays have lamellar structures and the concentration of amino acids between the layers favors their polymerization. This hypothesis has been confirmed by laboratory experiments, which have shown, for example, that peptide bonds can form between two amino acids trapped between the layers of a particular clay, montmorillonite, which thus catalyze the formation of RNA. This question of catalyzing the formation of primordial macromolecules is really central. In today's life, specialized proteins, enzymes, are playing this role, for example, by catalyzing the production of nucleic acids, DNA and RNA. But it is the DNA that contains the genetic information necessary to produce proteins... To get out of this vicious circle, the formation of the first macromolecules, or even of prebiotic molecules, must have been catalyzed in another way. It is very likely that clays and different minerals have played this role of heterogeneous catalysis. We have just seen this for RNA and clays, but it has also been shown that the iron oxides found in hydrothermal vents can catalyze the formation of ammonia from nitrogen and molecular hydrogen. The German chemist and jurist Günter Wächtershäuser has proposed that several of the steps leading to life could have taken place at hydrothermal vents. The synthesis of amino acids would have occurred deep in the earth's crust; these amino acids would then have been injected with hydrothermal fluids into colder waters, where lower temperatures and the presence of clay minerals and catalysis by iron and nickel surfaces would have favored the formation of peptides. The hydrothermal sites finally gather all the physico-chemical conditions favorable to the appearance of life; this hypothesis is particularly attractive because these sites of the primitive Earth, with its intense volcanic activity, must have presented on the surface many sites comparable to the current hydrothermal vents or to certain geyser fields.

5.7 The Formation of Cells

Compared to the difficult problem of the formation of proteins or DNA, that of cells seems to result quite naturally from the evolution of organic molecules, as soon as amphiphilic molecules such as lipids appear. In fact, lipid monolayers can spontaneously assemble into flat, then curved double structures, as shown in figure 5.8: they thus separate an inner and an outer medium, as in a living cell, and constitute filters capable of letting certain molecules through and retaining others. Laboratory experiments using amino acids extracted from the Murchison meteorite have seen more or less spherical, liposome-like vesicles developing rapidly in water. We are not far from the cell walls!

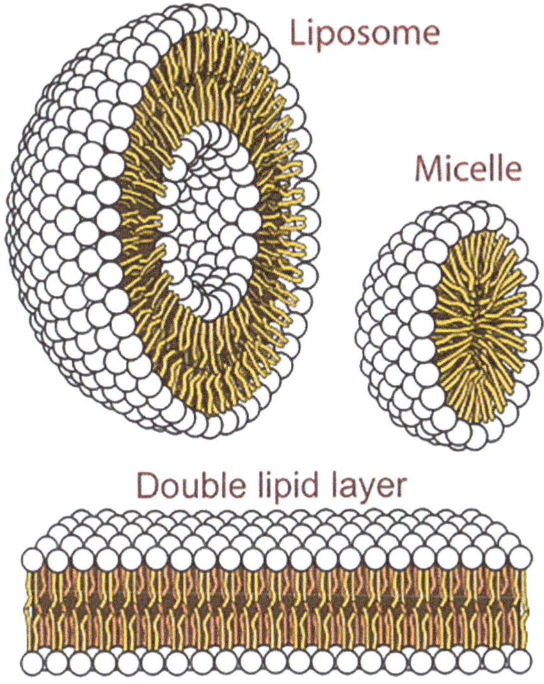

FIG. 5.8 — Lipid layer structure. The hydrophilic heads of the molecules are represented by white balls, their hydrophobic tails by yellow filaments. Monolayers have the ability to close in on themselves, forming micelles, which are frequent in colloidal solutions. Double layers can also close in on themselves, forming liposomes that can be artificially manufactured. Wikimedia Commons, Lady of Hats.

It should be noted that as early as 1924, Oparin, the Soviet biochemist at the origin of the primitive soup hypothesis, observed the formation of such droplets when proteins are added to water, under the name of *coacervates*.

We also need to understand not only how real cell walls can be formed but also how they can trap macromolecules such as proteins or DNA. Experiments have shown that if a mixture of vesicles and these macromolecules are subjected to cycles of hydration and dehydration, their entrapment can occur because bonds form to replace those left free by the departure of water. This is therefore a point in favor of the scenario that shows living organisms in periodically dried out ponds.

5.8   Metabolism and the Question of Energy

Terrestrial organisms derive their energy from complex molecules, through a series of biochemical reactions by which they are transformed into simpler molecules. In living cells, this energy storage is mainly through a chemical compound called adenosine triphosphate (ATP), which is capable of storing energy in a bond between a phosphate group and the adenosine diphosphate molecule that contains two of these groups. Energy is released when this bond is broken by hydrolysis.

ATP is only one of the many complex molecules that terrestrial life needs, and to build them it shows a prodigious ability to extract carbon, electrons and energy from various environments. So-called autotrophic organisms synthesize them from light or non-organic chemical elements present in their environment. Photo-autotrophs such as plants derive their energy from light and carbon from CO_2, chemo-lithotrophic life uses oxidation–reduction reactions of non-biological compounds such as minerals, and chemo-litho-autotrophic life uses CO_2 or other small carbonaceous compounds as a carbon source. Heterotrophic organisms, like humans, obtain carbon and energy by consuming autotrophs or other heterotrophs, or the organic molecules they produce.

The scenarios we have presented suggest that primitive living beings fed on the organic molecules present in their environment, whether they were brought from space or taken from hydrothermal sources. However, the materials brought from space were rapidly depleted, and in order to colonize the whole planet or at least its oceans, the evolution towards autotrophy proved to be a decisive advantage. Chemolithotrophy seems to have appeared very early, with reactions of oxidation of hydrogen by various carbon compounds such as carbon dioxide, producing methane and water; we then speak of methanogenesis, a character shared by many microorganisms considered primitive, the archaea. Hydrogen can also be oxidized in the presence of sulfur or iron, especially near hydrothermal springs, whose chimneys are often lined with pyrite (Fe_2S). The metabolism of sulfur in particular seems to be very old and is used by many thermophilic archaea, which draw their energy from the oxidation of hydrogen sulfide (H_2S) to transform mineral carbon into organic matter.

Primitive life developed in an Earth poor in atmospheric oxygen and the so-called oxygenic photosynthesis appeared later, between 1 and 2 billion years after the formation of the Earth. There is also a photosynthesis that does not produce oxygen, that of purple bacteria that oxidize hydrogen sulfide (H_2S) into elemental sulfur. Organisms that derive their energy from light and their carbon from CO_2 are photo-autotrophs, like plants, but there are also photo-heterotrophs. In fact, it seems that all possible combinations between these different sources of energy and carbon exist.

The variety of mechanisms that allow living organisms to exploit their environment testifies to the faculties of adaptation and evolution that have allowed living organisms to colonize the planet. Metabolism appears here to be intimately linked to the genetic code.

5.9   The Genetic Code

The genetic code allows the cells to reproduce identically while leaving a possibility of evolution. At the molecular level, this involves biomolecules capable of replicating, catalyzing, and also evolving without losing their ability to replicate. In today's world, this function is fulfilled by DNA, but it is extremely unlikely that the most primitive life was DNA-based. RNA appears to be a good candidate because it has certain catalytic properties. But was it the first molecule capable of replication, the "first replicator"? This is not certain, because RNA is easily destroyed and relatively unstable. Most researchers currently believe that RNA was preceded by another replicator, and various possibilities have been advanced: lipids, proteins, peptides... The synthesis of these first replicators, their formation by polymerization and even their replication had to be catalyzed or guided by a substrate, mineral substances or clays. Later, they would have evolved by selection of certain variants and would have detached themselves from their substrate.

A "RNA world", with a variety of proto-RNAs, would therefore have preceded the DNA-RNA-protein world we know. The various types of RNAs would have evolved under the selection pressure of the environment, with new variants and mutants more efficient or faster to reproduce appearing over time. Such directed RNA selection experiments have been conducted *in vitro*. The DNA then appears as a modified RNA, more efficient as a carrier of genetic information. What is the advantage of DNA over RNA? It is the stability of its double helix, much longer than RNA strands, allowing the constitution of large genomes, and also the possibility of evolution thanks to replication errors.

However, reproduction is only one of the characteristics of living beings, next to metabolism (energy) and compartmentalization (cell). What first appeared: reproduction, metabolism, the cell? Researchers are divided. For the "replication first" school, the first replicating molecules appeared first, then RNA, then the first cells. For the "metabolism first" school, to which the Oparine–Haldane hypothesis is related, it is abiotic organic chemistry that allowed the formation of macromolecules such as peptides and polysaccharides, then inside protocells, the first molecules capable of replicating. In any case, these steps cannot be very far apart in time, because protocells are necessary to separate living organisms from their environment. It is around the time when all these elements are in place that we can speak of the appearance of life.

5.10   The Ancestor of All Living Beings?

Reconstructing all the steps that led to the first microorganisms proves difficult, which is quite natural because even the smallest units of living organisms are extremely complex and result from evolution over millions or even billions of years (the ribosome, the tiny machine that in all living cells makes proteins, is made up of more than 100 000 atoms even in bacteria, and many more in eukaryotes). Another approach is more of an archaeological type, or even a police investigation: can we define a set of genes and biochemical characteristics that all living beings would share? Let us recall the theory of evolution and its characteristic evolutionary trees: evolution also leaves traces in the genes and nucleic acids of living beings. Thus, by comparing the ribosomal RNA sequences of a large number of species, scientists have been able to propose a "universal tree of life" with three branches: bacteria, archaea and eukaryotes. This tree corresponds to the two main types of cell organization, prokaryotes (bacteria and archaea, each with its own branch) and eukaryotes. Recall that prokaryotes have cells without a nucleus, unlike eukaryotes with their well-differentiated nucleus containing chromosomes. Archaea differ from bacteria in particular by the composition of their cell membrane.

The tree of life confirms at the level of their genes the existence of these three main branches of life, but above all their common origin, the "base" of the tree: LUCA, the "Last Universal Common Ancestor", or "last common ancestor". The age of LUCA is at least 3.5 billion years, probably a few hundred million years after the appearance of the first life forms (see chapter 6). In the tree of life, LUCA appears just before the great bifurcation between bacteria and archaea, and then between archaea and eukaryotes, which include fungi, plants and animals (figure 5.9).

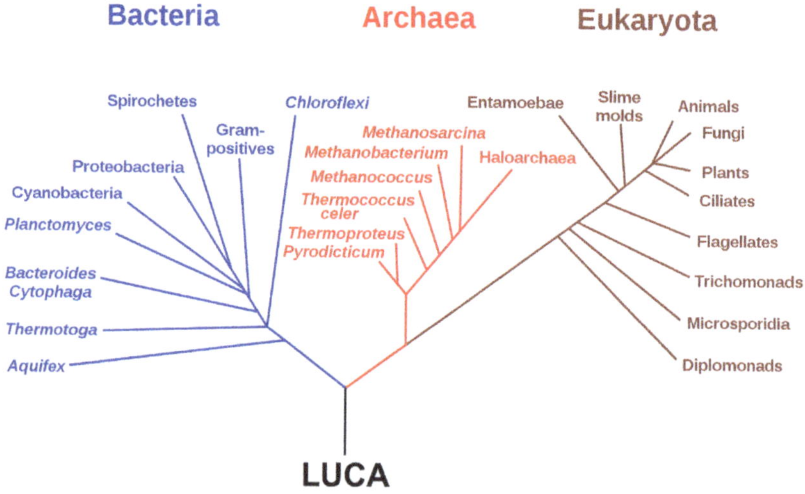

FIG. 5.9 – Diagram representing the tree of life, based on the Universal Tree of Life of Carl Woese (1928–2012), published in 1990. Wikimedia Commons.

It should be noted that LUCA was certainly not the only type of organism that lived in its time: it is the only one that had (long) offspring, bacteria, archaea and eukaryotes. Moreover, the genetic analysis of the tree of life shows that most organisms close to the origin of the branches, including LUCA, are extremophiles, thermophiles and hyper-thermophiles. They lived in very hot environments and did not use light as an energy source. Among them are methanogens and those who use the metabolism of sulfur, iron... They were, therefore, very similar organisms to those currently living in hot springs and hydrothermal vents.

However, this observation does not disqualify the ponds and shores discovered by the tides, and therefore the surface of the continents, as a place where life appeared: indeed, a very long time may have elapsed between the formation of the first pre-biotic molecules, that of the first replicator molecules, and LUCA. During this time, life was able to colonize the bottom of the oceans. The thermophilic nature of LUCA and the oldest organisms does not mean that the very first organisms were thermophilic either. Indeed, it is very possible that a major impact during the Late Heavy Bombardment (chapter 2) vaporized most of the oceans and destroyed all living things, except for the extremophiles that lived right next to the underwater chimneys, and which could later colonize the rest of the planet.

All this took time: 2.3 billion years elapsed between the appearance of the first bacteria and the first multicellular beings, and another 650 million years before the explosion of life in the Cambrian era. In particular, it was necessary to "invent" the chlorophyll of plants, which performs photosynthesis by absorbing CO_2 and producing oxygen, which now constitutes a significant part of the atmosphere: this happened about 500 million years after the first unicellular beings. On the Earth, evolution has been marked by climatic variations, continental drift, several episodes of extinction of many species, and most recently by the action of man. We will discuss this in more detail in the next chapter. There is no reason why evolution after the creation of the first cells should have a similar history on other planets that would eventually harbor life.

5.11   Life on Earth as a Model for Life on Other Planets?

To what extent can we rely on terrestrial life to extrapolate it to life on other planets? For example, can we imagine a biochemistry that is not carbon-based? Silicon has been often advanced as a possible alternative. Certainly, silicon shares certain characteristics with carbon, for example its tetravalence, but its chemistry is much less rich. Thus, among the molecules detected in the interstellar medium, only a few silicon-containing molecules and no chains are known, whereas most of the molecules contain carbon. Silicon can form polymers comparable to hydrocarbons, in particular silanes which are chains with Si–Si and Si–H single bonds, but this supposes a very particular environment: very poor in oxygen, water and carbon, negative temperatures, methane or some methanol-type solvent. In any other environment, the stable form of silicon will rather be silicates. Such conditions do not seem to be frequent neither in the solar system nor in the known exoplanets; the only example that could approach them, although rich in carbon, would be Titan, Saturn's satellite, which would certainly be interesting to explore in depth.

What about water as the universal solvent for life? After all, our idea of a habitable planet is univocally based on the presence of liquid water for a sufficient length of time for the appearance of life; if life can use other solvents, this could significantly expand the range of potentially habitable worlds. For example, for planets colder than Earth, liquid ammonia might be a good polar solvent; but in an environment such as Titan's, non-polar organic solvents such as methane and ethane should be considered. In any case, the fact that the solvent is not water would induce a biochemistry very different from that which we know on Earth, and in fact hard to imagine in the absence of observational or experimental elements, which makes it difficult to define criteria for detecting life based on such solvents. Moreover, water finally appears to be the most "universal" solvent in the sense that all the others are adapted to a rather restricted range of physico-chemical conditions.

Do water as a solvent and carbon as a basic chemical element imply a life based on the 20 amino acids we know and nucleic acids, DNA and RNA, using the 5 nitrogenous bases of terrestrial life, AGCT for DNA, AGCU for RNA? Nothing is less sure. Indeed, what counts for the functioning of molecules of the living is not so much their exact chemical formula but their function. For example, the catalytic properties of proteins are intimately linked to their three-dimensional configuration, itself linked to the interactions between the dipoles that make up the protein chain and the possibility of creating hydrogen bonds. But this does not imply that only a combination of the 20 biological amino acids will do the trick. It is moreover one of the objects of study of the sub-discipline called xenobiology to explore possible forms of biology that would differ from the canonical system based on DNA, RNA and 20 amino acids. For example, it has been shown that an extended genetic "alphabet" could be constructed using up to 8 nitrogenous bases to make an AXN, a xeno-nucleic acid: the usual 4 bases plus 4 others. It is also possible to modify the DNA backbone chain by replacing the nucleotide sugar (deoxyribose in the case of DNA) with another sugar. In most cases, the functions of the modified molecules are preserved. Other researchers have modified bacteria to make them function by replacing one of the nitrogen bases, thymine, with another compound; after a few thousand generations, the bacteria had become able to do without thymine.

These experiments, beyond their science-fiction aspect, bear witness to the process of evolution and selection that led to the biology of terrestrial life: the final result was not dictated in advance in all its details. This observation should lead us to caution when we try to detect life on other planets: if in all likelihood this life will be based on carbon and water and will manifest itself through an organization and phenomena out of balance, it is perhaps not with a DNA chip that it will be detected.

5.12   The Beginnings of Life on Earth

When did life appear on Earth? It seems to have been more than 4 billion years ago: at least, this is indicated by the recent discovery of carbon of biological origin in zircons dated 4.1 billion years ago. It is also about the time of the end of the "Late Heavy Bombardment". Life may have appeared earlier, but we do not have rocks old enough to have kept track of it. Fossilized remains of the first organisms date back at least 3.5 billion years, with stromatolites in particular testifying to bacterial communities: there is no doubt that life was present at that time. Photosynthesis appeared at the earliest 3.2 billion years ago, and the first cells with eukaryotic nuclei 2 billion years ago. Throughout this period, the inhabitants of the Earth were microorganisms living in the seas and oceans.

What happened before the first organisms emerged that we can recognize as living and that have some kinship with archaea and bacteria is still poorly understood. The necessary steps have been identified and some of them demonstrated in the laboratory. The first was the accumulation of the basic building blocks: simple prebiotic molecules, which may have come from space and/or been synthesized from hydrothermal sources. Several steps are then necessary and the order in which they occur is not clear: the synthesis of molecules capable of constituting protocell membranes, the construction of increasingly complex molecules capable of storing information and then translating it, and the selection of catalysts that make all these processes possible. It is very likely that these steps involved mineral surfaces in the primitive ocean or clays. Where did these steps take place? For many of the reasons we have mentioned, hydrothermal vents are very plausible candidates. But one can also envisage with good arguments shores periodically discovered by the tide, temporary ponds or even micro channels in water ice that would undergo cycles of freezing and thawing.

Although life may have appeared relatively quickly on Earth, this must probably be measured in millions or even tens of millions of years. It is therefore not very surprising that we have difficulty reproducing in the laboratory, in a few years, the phenomena that led to the first living or "semi-living" beings, to use Haldane's words. It is also very likely that these distant ancestors were very different from today's living beings, hence our difficulties in imagining and studying them.

5.13   Life on Exoplanets

Our knowledge of the appearance of life on Earth is still so uncertain that any detection of past or present life on a planet in the Solar system would give us valuable information. For example, researchers believe that during its first billion years, the planet Mars may have had an ocean. This is about the same time that life appeared on Earth, and the conditions on the two planets must have been quite similar at that time. All traces of this remote period have disappeared on Earth, but Mars, which has neither plate tectonics nor water erosion, may have recorded fossil traces of primitive microorganisms.

To look for life elsewhere, however, indications of where it has appeared on Earth would be very useful. If life was born in puddles sown with extraterrestrial materials, it may be useless to look for it under the ice floes of the ice moons of the giant planets. If it was born in tide-covered and tide-discovered foreshores, we should be interested in exoplanets that have a large satellite; in the Solar system, there would then be few candidates outside the Earth, and perhaps Titan. And if it is near hydrothermal springs, then the icy satellites of giant planets become prime targets, especially Europa where it is thought that the ocean under the icy surface is in contact with rocky ground.

However, let us remember that for more than 3 billion years, the only inhabitants of the Earth were microorganisms living in the oceans. Direct observation of these microorganisms on an exoplanet is obviously impossible. But a planet that is home to life, a phenomenon out of equilibrium by nature, is perhaps profoundly transformed, and even underwater life could result in a drastic modification of the planet's atmosphere. This is what we will discover in the next chapter.

Chapter 6

The Development of Life on Earth

Let us go back to planet Earth, whose climate archives we can explore to understand the past evolution of the planet and try to trace the scenario that led to the appearance and then the development of life.

First of all, what means do we have to go back in time? A first tool is the measurement of the chemical composition, elemental and isotopic, carried out in rocks whose age can be evaluated by measuring the radioactive decay of certain elements. Core drilling of the ice caps (figure 6.1) gives us information on the evolution of temperature over the last million years. Another tool is provided by measurements of the magnetic field that has reversed over time and whose trace, in some rocks, allows us to determine their age. To go further back in time, climatologists use radiogenic elements of long period (such as the uranium-strontium couple), and geological indicators, such as traces left by glaciations, the study of plant or animal fossils, or liquid or gaseous inclusions in ancient sediments. We can thus estimate the average temperature over the last hundreds of millions of years, and even as far back as the Archean, more than two billion years ago.

Let us go back to the origin of the terrestrial planets. We have seen that the atmospheric composition of Venus and Mars was dominated by carbon dioxide, with a low percentage of molecular nitrogen. Located between Venus and Mars, the primitive Earth must have had the same atmospheric composition. Like its neighbors, the Earth had to undergo a massive meteorite bombardment, coming in particular from the outer Solar System, itself the origin of part of the water in the Earth's oceans.

6.1 The Paradox of the "Young Sun"

In the case of the Earth, we will find a problem that we have already encountered in the case of the primitive atmosphere of Mars: it is the paradox of the "young Sun". We have seen that some four billion years ago, the luminosity of the Sun was about 30% weaker than today. This luminosity was then insufficient for the planet's

DOI: 10.1051/978-2-7598-2563-9.c006

FIG. 6.1 – An ice core from the EPICA drill hole in Antarctica. The European EPICA project has made it possible to trace the history of the Antarctic climate back 800 000 years. The parameters measured are the amount of carbon dioxide and the amount of methane; we can also have an estimate of the temperature. ©CNRS/INSU, Augustin Laurent.

equilibrium temperature to be above 0°C: the Earth should have been in total glaciation. As the albedo of ice is very high, the solar radiation should have been strongly reflected, further lowering the equilibrium temperature and making any warming mechanism impossible in principle.

What is the mechanism that allowed the Earth to come out of the state of permanent glaciation? Climate specialists agree on the greenhouse effect generated by carbon dioxide injected in the Earth's atmosphere by volcanism. Indeed, at the beginning of its history, the Earth's internal energy (mainly the energy released by the radioactive elements present in the mantle) was greater than today, resulting in a much more intense volcanic activity. Moreover, in the first billion years, in the absence of continents, carbon dioxide must have accumulated in the atmosphere as a result of successive volcanic episodes, since its sequestration into calcium carbonate could not take place in the absence of bedrock.

However, it is not certain that carbon dioxide alone could have generated a greenhouse effect sufficient to break the cycle of the initial glaciation. Another gas may have contributed to the greenhouse effect: methane, which is also very effective from this point of view. Absent at the very beginning of the Earth's history, methane must have been generated by the methanogenic archaea, traces of which have been discovered dating back 3.5 billion years; its concentration probably increased until the appearance of oxygen in the atmosphere (what is called the Great Oxidation Event, or GOE; we will come back to this later), some 2.5 billion years ago.

6.2 The Major Stages in the Evolution of the Earth's Climate

Geologists have divided the Earth's history into four major periods or "eons": Hadean (4.6–4.0 billion years), Archean (4.0–2.4 billion years), Proterozoic (2.4 billion years–540 million years) and Phanerozoic (540 million years to the present). Life appeared in the underwater environment at the end of the Hadean and developed during the Archean. During the Proterozoic period, the first oxygen-producing bacteria appeared, still in the oceans. Finally, the Phanerozoic, which began 540 million years ago, saw the development of life on the surface of the continents (table 6.1).

6.2.1 From Hadean to Archean

The oldest certain traces of life come from underwater bacterial microfossils dating from the end of the Hadean or the beginning of the Archean. These are mainly stromatolites, calcareous concretions built by single-cell algae, discovered in particular in Australia (figure 6.2); other forms of microorganisms are also found near hydrothermal springs, also more than 3.7 billion years old. In addition, isotopic measurements carried out in Australia have made it possible to characterize carbon whose origin could be biogenic in zircons, crystals formed from zirconium silicates dating back more than 4 billion years, embedded in magmatic or metamorphic rocks. It is therefore not impossible that the first life forms appeared before the Late Heavy Bombardment, which is about 3.7 billion years ago. Did they irretrievably disappear, and did life then make a new beginning after this cataclysmic episode, or did some of them survive? The question remains open.

As soon as the oceans appeared and a silicate soil was present (as evidenced by the Jack Hills zircons), towards the end of the Hadean, a major reaction took place between carbon dioxide and the soil, which led to the sequestration of CO_2 in the rocks in the form of limestone (see box 6.1 at the end of this chapter). This is a two-stage reaction that sees first, in the presence of liquid water, the formation of bicarbonate HCO_3^- from magnesium silicate and carbon dioxide, and then the formation of calcium carbonate from bicarbonate. The net balance is the sequestration of carbon dioxide into limestone. This reaction became effective in the oceans when

Tab. 6.1 – The Earth's major geological periods.

Date (million years)	Eon	Main features	Atmospheric composition	Evolution of life
4600–4000	Hadean	4600 My: Formation of the Earth Primordial atmosphere: CO_2, N_2, $+H_2$, H_2O	CO_2, N_2	Before 4100: No evidence of life 4100: Presence of liquid water and first traces of life (zircons)
4000–2400	Archean	3700: *Late heavy bombardment* → Appearance of water in the atmosphere 2900: Formation of continents	Decrease of CO_2 Appearance of CH_4	3700: Appearance of the first methanogenic bacteria 3000: Appearance of the first oxygen-producing bacteria
2400–540	Proterozoic	2400: Planetary glaciation 720: Planetary glaciation 635: Planetary glaciation	2400: Appearance then increase of O_2 (*Great Oxidation Event*) Decrease of CH_4	Life thrives in the oceans 2000: Appearance of eukaryotes (first cells with a nucleus) 1200–540: Appearance of the first multicellular organisms
540–Present	Phanerozoic	540: Cambrian biological explosion 320–270: Permio-carboniferous glaciation 65: Meteorite fall at Chicxulub 34: Antarctic glaciation 2.7: Greenland glaciation	540: Emergence of a stratospheric ozone layer 50: CO_2 reduction 0.001: Beginning of the industrial era and anthropogenic CO_2 increase	360: Plants grow on continents 250: Permo-Trias biological extinction 200: Triassic-Jurassic biological extinction 65: Cretaceous-Tertiary transition and mammalian development

FIG. 6.2 – A field of stromatolites in Shark Bay, Australia. Stromatolites are calcareous laminar structures built by bacteria. The oldest are 3.7 billion years old. Most of the stromatolites found today are fossils. Those from Shark Bay are among the few still in development. Wikipedia/Paul Harrison.

the oceanic crust was sufficiently cooled to allow the recycling of carbonate sediments by subduction and the sequestration of carbon underground, which may have taken place some 4 billion years ago. The net effect of this reaction was a decrease in the CO_2 content of the atmosphere throughout the Hadean and Archean periods. Later, the CO_2 sequestration mechanism also occurred in the emerged soils, when heavy rains watered the silicate soils of the continents.

During the Archean, it seems that some of the first living organisms were cyanobacteria, which have the property of producing methane; they are the ones that are at the origin of stromatolites. Methane, whose concentration in the atmosphere then increased, played an important role in the evolution of the Earth's climate because it contributed, along with carbon dioxide, to the greenhouse effect that allowed the Earth to escape total glaciation. At the same time, the first microorganisms consuming carbon dioxide appeared, and the atmospheric CO_2 decreased. As for the temperature of the oceans at the time of the Archean, according to isotopic measurements of oxygen and silicon made on Archean rocks, it appears to have been much higher than today (between 50 and 80°C according to various estimates). This unexpectedly high temperature seems to be in contradiction with the scenario described above, which involved successive periods of glaciations imposed in the Hadean by the low luminosity of the young Sun. This paradox is currently unexplained.

6.2.2 *From Archean to Proterozoic: The Great Oxidation Event*

At the end of the Archean, the first oxygen-producing bacteria appeared; their effect was considerable because oxygen reacts quickly with methane to destroy it, producing carbon dioxide and water. This led to a rapid decrease in the atmospheric concentration of methane, and the transition from Archean to Proterozoic, 2.4 billion years ago, was marked by the rapid increase of molecular oxygen in the atmosphere (figure 6.3). This is due to oxygenic photosynthesis that, in the presence of light and water, produces oxygen and organic matter. Initiated by cyanobacteria, the production of oxygen had the first effect of oxidizing the iron present in marine soils.

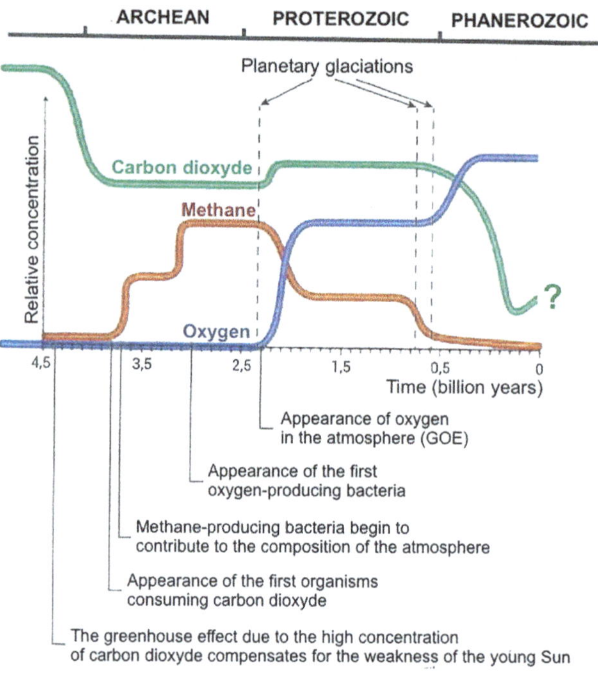

FIG. 6.3 – Evolution of the Earth's atmospheric composition as a function of time. The atmosphere, initially dominated by carbon dioxide, was enriched in methane with the appearance of cyanobacteria 3.7 billion years ago, then in molecular oxygen during the great oxidation event 2.4 billion years ago. Adapted from G. Ramstein, LSCE, *Reflets de la Physique* 55, 6 (2017).

The beginning of the Proterozoic, 2.4 billion years ago, also saw the first global planetary glaciation of which we have a trace, known as the Huronian. It could have been the consequence of the decrease in atmospheric methane associated with the surge of molecular oxygen (which is not a greenhouse gas) and also, perhaps, large meteoritic falls or very strong volcanic eruptions; these phenomena, by projecting into the stratosphere large quantities of dust that absorbs solar energy, lead to a cooling of the atmosphere. Little is known about this episode because our knowledge of the distribution of oceans and land masses does not extend beyond 1.5 billion years. However, there are craters known to have originated in major meteorite impacts: the Vredefort Crater in South Africa, with a diameter of 300 km and

2 billion years old, and the Sudbury Crater in Canada, with a diameter of 250 km and 1.85 billion years old. Similar events certainly have taken place in the older past, but erosion mechanisms and plate tectonics have erased their traces.

How did the Earth come out of the first Proterozoic glaciation? When the surface was completely icy, the sequestration of carbon dioxide in the soils could no longer take place, so its concentration in the atmosphere increased steadily due to volcanism, hence the greenhouse effect. It is also possible that violent volcanic eruptions injected large quantities of carbon dioxide into the atmosphere. In fact, volcanic activity can have opposite effects: if it injects particles into the atmosphere, it decreases the solar flux received at the Earth's surface and results in a cooling of the atmosphere; if it injects carbon dioxide in large quantities, it increases the greenhouse effect and warms it. Everything therefore depends on the type of volcanism associated with the period in question, about which we unfortunately have no information.

The second global glaciation, which occurred some 600–700 million years ago, is, on the other hand, better known because we can trace the distribution and evolution of the continental masses at that time. It could be due to the particular configuration of the continents. 800 million years ago, a super-continent, Rodinia, was located at low latitude; because of its large size, it was little watered, except in the periphery. Over the next tens of millions of years, it broke up into small continental plates, while remaining at the same latitudes. Subjected to heavy precipitation, these plates underwent a silicate alteration that sequestered some atmospheric carbon dioxide, thus causing a drop in temperature. The Earth came out of this period, once again, thanks to the accumulation of atmospheric carbon dioxide (which no longer interacted with the frozen surface), probably produced by intense volcanic activity, and the resulting greenhouse effect.

6.2.3 *The Phanerozoic: Life on the Continents*

As soon as it was present in the Earth's atmosphere, molecular oxygen was subjected to solar ultraviolet radiation. This resulted in the formation of a stratospheric ozone layer, the effect of which is crucial for the development of life: by absorbing solar ultraviolet radiation, the ozone layer protects the surface of the continents from its harmful effects and makes the development of complex organic molecules possible. The entry into the Phanerozoic era 580 million years ago thus marked the development of life on the surface of continents.

Climate change is therefore strongly linked to the tectonic activity of the planet, whether it be continental drift or the elevation of mountainous relief. The same applies to the carbon cycle and the atmospheric concentration of carbon dioxide. During the Phanerozoic, between 300 million and a few million years ago, ice ages became rarer and the average temperature was about ten degrees higher than today. The secondary era (Mezozoic) corresponds to the maximum level of water rise (several hundred meters above the current level) and the formation of large sedimentary basins, such as the Paris Basin. At the same time, the Atlantic Ocean opened up as well as the Indian Ocean. The Carboniferous period, between 360 and

295 million years ago, saw the appearance of ferns, horse-tails and giant plants as temperatures rose. Their decomposition in swampy areas led to the formation of large fossil reserves of carbon, in the form of coal, gas and oil. After the appearance of fish in the oceans 540 million years ago, the animal kingdom saw the successive appearance of amphibians, insects and reptiles. The Mesozoic era, between 300 and 65 million years ago, was the reign of dinosaurs and birds (figure 6.4).

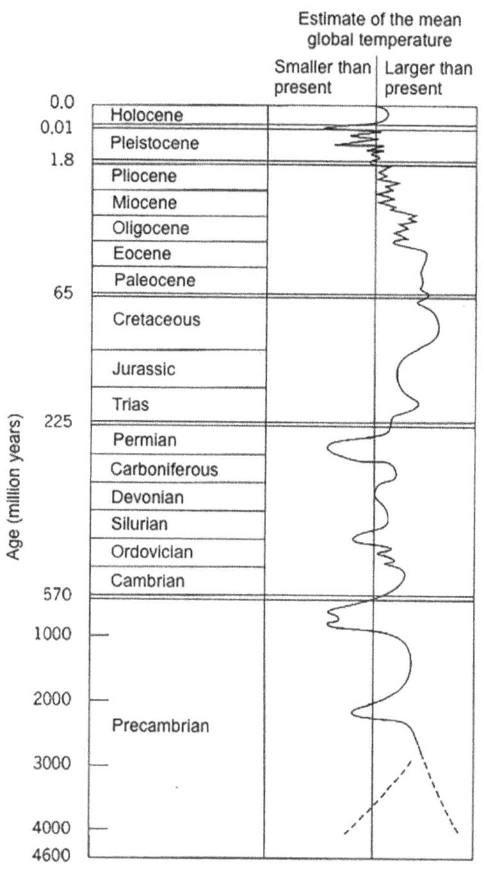

FIG. 6.4 – Evolution of the Earth's average global temperature as a function of time, evaluated from geological indicators. A maximum appears during the Jurassic and Cretaceous periods, about 100 million years ago. Adapted from S. Joussaume, *Climat d'hier à aujourd'hui*, Belin.

The Tertiary Era, which began 65 million years ago, is characterized by the disappearance of dinosaurs and many other animal species. This ecological upheaval is attributed to the fall of a giant meteorite, about ten kilometers in diameter, in Chicxulub, on the edge of the Gulf of Mexico; the phenomenon may have been amplified by intense volcanic activity in the Dekkan dating from the same period, which could have been triggered by the fall. This event had considerable importance since the extinction of the dinosaurs allowed the explosion of mammals with, in the end, the appearance of human beings.

It is possible to measure the evolution of carbon dioxide content since the beginning of the Tertiary era using various paleo-climatic indicators (phytoplankton, elemental abundances). These measurements show that the atmospheric carbon dioxide content has dropped by a factor of 3 between the beginning of the tertiary era and the end of the pre-industrial era. The reasons for this fall remain poorly explained. About 30 million years ago, the Antarctic continent became covered with ice, while the glaciation of Greenland occurred much later, about 3 million years ago. During the Quaternary Era, periods of glaciation followed one another, with ice caps covering all of Canada, as well as northern Asia and Europe. The work of the Yugoslav mathematician Milutin Milankovitch (1879–1958) showed a relationship between the dates of the appearance of periods of glaciation and the amount of solar flux received at the surface. This varies according to several factors: the eccentricity of the Earth's orbit introduces a periodicity of about 100 000 years, while variations in the obliquity of the Earth relative to the ecliptic introduce a period of 43 000 years; finally, the precession of the Earth's axis introduces two other modulations, of periods 24 000 and 19 000, respectively (figure 6.5). Broadly speaking, during the Quaternary, the evolution as a function of time of the carbon dioxide rate (and thus of the mean temperature) follows the received solar flux at 65° north latitude. The last ice age took place 21 000 years ago; sea level fell by more than 100 m from the present level, significantly modifying the contour of the continents. Thus, the English Channel did not exist, as France and England were part of the same continent. This context allowed Homo Sapiens to conquer the whole of Eurasia.

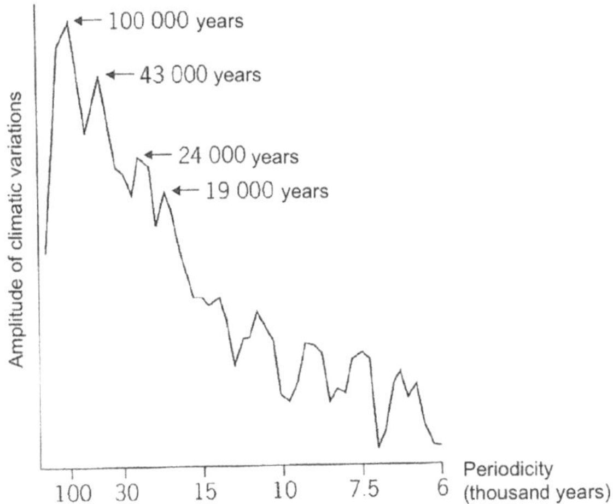

FIG. 6.5 – Evolution of climatic variations measured over the last 500 000 years from isotopic contents measured in marine sediments of the Indian Ocean. The spectral analysis of these variations reveals several periodicities that correspond to those of the Earth's orbital parameters, thus providing confirmation of Milankovitch's theory. Adapted from S. Joussaume, *Climat d'hier à aujourd'hui*, Belin.

However, as we move closer to the present time and the number and accuracy of measurements increase, the situation becomes more complicated. For example, 7000 years ago the Earth experienced a maximum temperature more than 2°C higher than the current average. Rainfall was then abundant and the entire Sahara was covered with green meadows. This episode was followed some 5000 years ago by a cooling that led to the desertification of many continents. Between the 10th and 13th centuries AD a "medieval climatic optimum" appeared, which saw the colonization of the green meadows of Greenland (hence its name) by the descendants of the Vikings coming from Norway. In a more recent past, a few centuries ago (between 1440 and 1880), Europe again experienced a "Little Ice Age" with a notable advance of the glaciers. It is followed by the current warming, due in large part to human activity, which we will discuss.

6.3   What Future for the Earth's Atmosphere?

In the 19th century, at the end of the pre-industrial era, the concentration of CO_2 in the atmosphere was 280 ppmv (parts per million by volume), and the average temperature was 0.6°C lower than in 2000. In 2019, the CO_2 level had reached 410 ppmv and the average temperature was 0.2°C higher than in 2000. These parameters are precisely measured thanks to drillings carried out in Antarctica to a depth of 3 km, which analyze, at different depths, the isotopic ratios of the elements contained in the ice. At the same time, the sea level rose by 20 cm. The results are indisputable: we are witnessing a runaway greenhouse effect fueled by carbon dioxide. What is more, given the very long lifetime of carbon dioxide in the Earth's atmosphere (more than 200 years), any increase in the rate of CO_2 will be perpetuated in the centuries to come: even if we were able to stabilize it today, it would not decrease for another two centuries.

What does the future hold? Global climate models that attempt to extrapolate these data into the future are, of course, somewhat uncertain, but their conclusions are worrisome. According to the latest IPCC (Intergovernmental Panel on Climate Change) report, sea levels could rise by 30–90 cm by the end of the 21st century, with catastrophic consequences for a very large number of people living near the coasts. In this increase, the melting of the ice caps could become the predominant factor in the future, ahead of the expansion of the oceans due to warming and the contribution of glaciers, whose effects dominate today.

In conclusion, it is now accepted that, following the advent of the industrial era and population growth, human activity today plays a decisive role in the climate, the effect of which accelerates with time. Although it is difficult to accurately predict the extent of the phenomenon, given the number and complexity of the parameters involved, we must take into account the possible risk of an irreversible runaway of the greenhouse effect. We must be aware of this risk on a planetary scale, and implement policies for energy saving, research into renewable energies and the protection of biodiversity.

Another remark is obvious, with regard to the evolution of the Earth's climate over the course of its history: there is a close correlation between the appearance of life in its various forms and the evolution of the Earth's climate. The appearance of the first methanogenic cyanobacteria was responsible for the increase of atmospheric methane; the formation of the continents, some 3 billion years ago, made it possible to trap atmospheric carbon dioxide in the soil in the form of carbonates. Later, the development of the first oxygen-producing bacteria led to the appearance of molecular oxygen in the atmosphere; then, the formation of a stratospheric ozone layer, product of the photolysis of oxygen, produced a protective screen that allowed life to conquer the continents.

Catastrophic events have played a major role in the evolution of life forms. Volcanic activity must have been a major element, by injecting into the atmosphere a quantity of carbon dioxide sufficient to reactivate the greenhouse effect in periods of glaciation, but also, conversely, by injecting solid particles into the stratosphere, capable of blocking solar radiation and causing periods of cooling. Finally, major meteoritic impacts have also played a decisive role in the evolution of life.

The Earth has experienced several mass extinctions during its history, identified from fossil records, the cause of which is often poorly known. Thus, the Triassic–Jurassic extinction, dating back some 200 million years, saw the disappearance of 75% of marine species and 35% of animal species. We have seen that, closer to us, 65 million years ago, the meteorite impact of the Chicxulub and its probable consequences on volcanism in the Dekkan led to the disappearance of dinosaurs and many other species. Without this ecological catastrophe, providential for the development of mammals and finally for the appearance of human beings, we would not be here to talk about it. Humans are presently responsible for another extinction, through the overexploitation of resources, pollution of the environment and climate change.

Finally, astronomical factors, linked to the orbital parameters of the Earth and the Moon, intervene in a decisive way on climatic variations on geological scales. We have seen that during the Quaternary, the temporal evolution of the periods of glaciation is closely correlated to the Milankovitch cycles, which involve the periodicities of the different orbital parameters of the Earth. But that is not all: the presence of the Moon, very soon after the formation of the planet, played a role in stabilizing its obliquity. Indeed, numerical simulations carried out by a team from Paris Observatory have shown that, in the absence of a sufficiently massive satellite, the obliquity of the telluric planets would oscillate with a period of about 20 million years between a value close to zero and a maximum of up to 60°; this is what happened to the planet Mars, with important consequences on the latitudinal distribution of its glaciers (see chapter 5). In the case of the Earth, calculations show that the presence of the Moon has the effect of stabilizing the obliquity of its axis, with important implications for the evolution of its climate.

6.4   What Lessons for Exobiology?

From the history of the development of life on Earth, can we extract some general remarks useful in the search for extraterrestrial life? While obviously refraining from generalizing from a single example, we can try to define some characteristics.

Let us first note that life seems to have appeared relatively early in the history of the planet, during the first billion years; it was probably born in an aqueous medium, in the form of microorganisms. It developed very slowly: unicellular organisms were the only ones for more than two billion years, until the beginning of the secondary era, less than 600 million years ago. If life is present on another planet, it is likely that it exists mainly in the form of microorganisms, because due to its high surface/volume ratio, a small cell is better suited for the exchange of matter and energy with the outside world. Even today, microorganisms are an important part of Earth's biomass.

Later, following the development of life on the continents, very rapid evolutions alternated with great extinctions due to external factors (falling meteorites and/or giant volcanic eruptions). These evolutions result in an increasing complexity of organisms which is globally accompanied by an increase in the size of living beings, going from the microscopic to the metric scale. It can also be noted that variations of external conditions (temperature, humidity) led, by adaptation of species to their environment, to an enrichment of biodiversity.

A final observation: once life appeared on Earth, it proved capable of adapting to the most extreme conditions. Indeed, microbial colonies have been discovered in extremely hostile environments of temperature, pressure, acidity or radiation. While we still do not know how to reproduce in the laboratory the passage from the non-living to the living, it must be recognized that when life appeared in a given environment, it is very difficult to make it disappear.

Box 6.1 Carbon dioxide sequestration in the oceans. The gradual cooling of the young Earth, during the Hadean or early Archean, allowed the condensation of water present in the atmosphere and the formation of oceans. The atmosphere then became dominated by carbon dioxide, whose pressure, depending on the models, could reach several tens or even a hundred bars. How then could we escape the runaway greenhouse effect of which the planet Venus was the victim? The presence of liquid oceans saved planet Earth from this fate. The commonly accepted pattern is as follows. First, in the presence of liquid water, magnesium silicate reacts with dissolved carbon dioxide in the ocean in the following reaction to form bicarbonate HCO_3^-:

$$MgSiO_3 + 2CO_2 + H_2O \rightarrow Mg^{++} + 2HCO_3^- + SiO_2$$

The bicarbonate then reacts to form calcium carbonate, which forms a calcareous crust:

$$2HCO_3^- + Ca^{++} \rightarrow CaCO_3 + CO_2 + H_2O$$

The result of the two reactions is that one CO_2 molecule was trapped as limestone. This mechanism allows the reduction of CO_2 in the atmosphere and the limitation of the greenhouse effect.

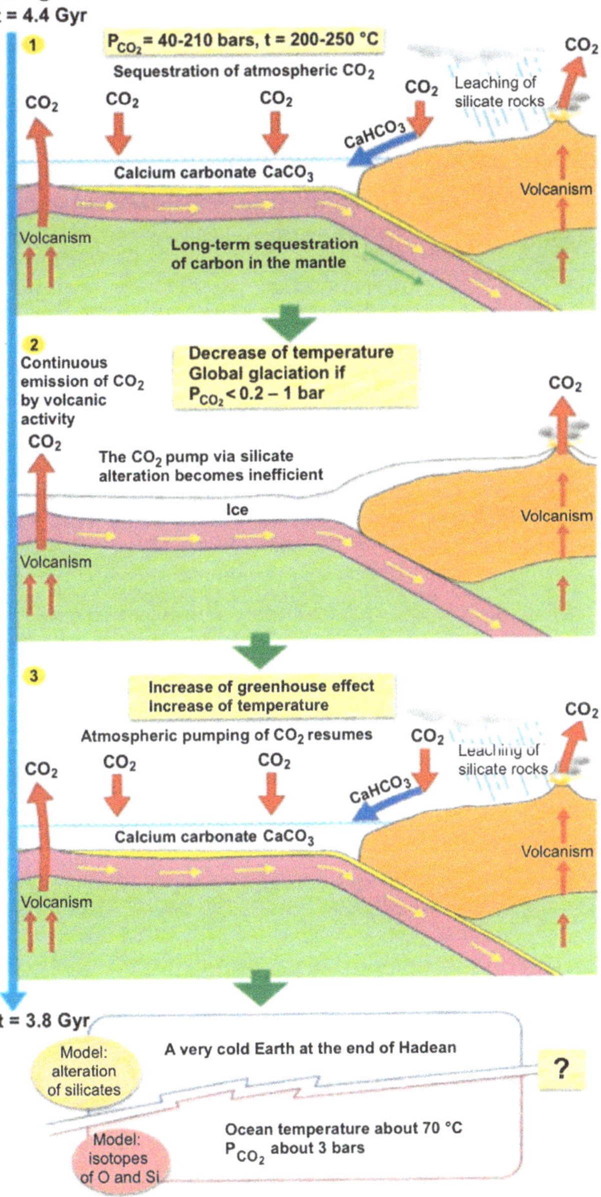

FIG. B.6.1 – The carbon dioxide cycle induced by the formation of limestone. Adapted from Gargaud *et al.*, *Le Soleil, la Terre, la Vie: la quête des origines*, Belin, Paris.

This mechanism can also work on land if the silicates on its surface are subjected to heavy rainfall. We can thus imagine, at the beginning of the Earth's history (Hadean and Archean), the following cycle for CO_2: (1) Carbon dioxide, initially very abundant and outgassed episodically by volcanism, was trapped in the form of limestone and the carbon was sequestered in the mantle; (2) The decrease in CO_2 was sufficiently important to trigger a global glaciation and the Earth was covered with ice; (3) Carbon dioxide, always emitted by volcanism, accumulated in the atmosphere but no longer interacted with the silicate soil due to the presence of ice; the temperature rose again until it caused the ice to melt, and we were back to case 1. This carbon cycle has certainly played a role in the successive glaciations throughout the Earth's history.

Chapter 7

Life in the Solar System?

From our journey into the past of the Earth's atmosphere, we can retain several essential, even indispensable, elements that led to the advent and development of life as we know it. These are, of course, the presence of liquid water, but also the presence of "nutrients" (carbon, hydrogen, nitrogen, oxygen, phosphorus and sulfur) and finally a source of energy. This energy can come from solar radiation, but also from hydrothermal heat sources or chemical reactions.

7.1   The Habitability Zone in the Solar System

In the case of the Earth, the presence of liquid oceans has been essential for the appearance of life. From this simple observation was born the concept of "habitable zone" introduced in the 1950s by several researchers. The habitable zone (HZ) around a star – starting with the Sun – describes the region where the temperature is such that liquid water can be found on the surface of a planet. In the case of the Earth, for an atmospheric pressure of 1 bar, the temperature range is between 0 and 100°C. The phase diagram of water (figure 7.1) shows that, while the upper limit of this range depends on the ambient pressure, the lower limit is always very close to 0°C. From this diagram we can isolate two regions of potential interest for habitability. The first is the region close to the triple point, at moderate pressures and temperatures, in the range 0–100°C. It corresponds to the atmospheric conditions of terrestrial planets and, by extrapolation, rocky exoplanets. The second region is at higher temperatures and pressures, up to the critical point ($P = 221$ bars, $T = 374$°C); it is the region of the liquid oceans that one expects to find inside the outer satellites of the giant planets; we will evoke them at the end of this chapter.

The concept of habitable zone as defined here concerns the first category: we consider rocky planets in orbit around a star whose radiation is such that its surface can receive and store liquid water (figure 7.2). However, it should be kept in mind that the second category, without entering into the definition of the "habitable zone", could also, *a priori*, host prebiotic chemistry, or even living beings; the study

DOI: 10.1051/978-2-7598-2563-9.c007

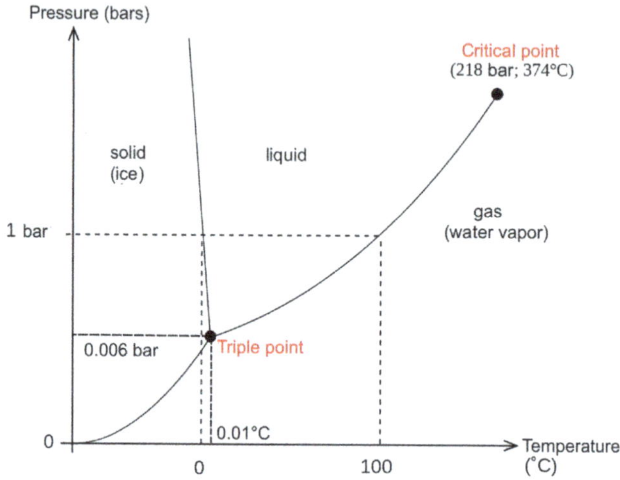

FIG. 7.1 – The phase diagram of water. The line separating the solid phase from the liquid phase is practically vertical: the transition between the solid and liquid forms of water occurs at a temperature close to 0°C over a very wide range of pressures. Adapted from Wikipedia, Creative Commons.

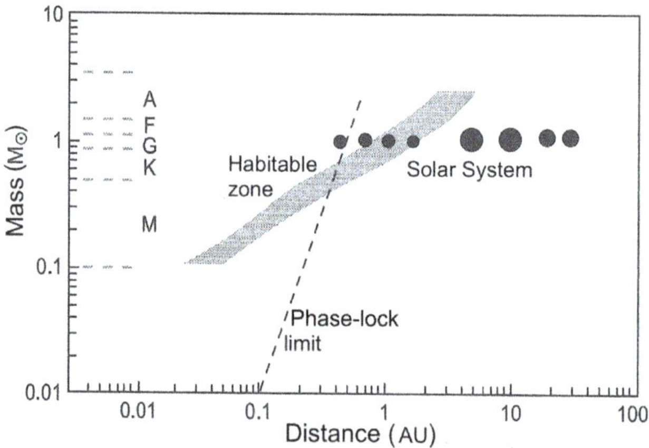

FIG. 7.2 – The habitability zone (or habitable zone) of a main-sequence star, as a function of its mass and spectral type (left). The more massive and brighter the star is, the further away the habitable zone is. In the case of the Sun (spectral type G2V), the habitable zone includes the Earth and Mars but not Venus. From Catling and Kasting, *Atmospheric evolution of inhabitable and lifeless worlds*, Cambridge University Press, 2017.

of potential habitats linked to the satellites Europa of Jupiter, and Enceladus and Titan of Saturn, is currently the subject of active research. If the concept of habitable zone, as defined here, has focused on planets with a surface, it is in the perspective of a future search for life on these objects: this should *a priori* be easier (or less difficult!) for life on the surface than for underwater life hidden under a layer of ice.

What are the boundaries of the habitable zone in the case of the Solar System? The internal limit is set by the vaporization of liquid water in the stratosphere as a result of photodissociation of water and escape of hydrogen. According to the most recent climate models, this limit is 0.95 au. The outer limit is more difficult to quantify. What happens when we move away from the Sun? When the temperature decreases, the continents become covered with ice and the carbon dioxide emitted by volcanoes accumulates in the atmosphere, increasing the greenhouse effect and slowing down the cooling process. There comes a time when clouds of solid carbon dioxide form in sufficient quantities to inhibit the greenhouse effect. Researchers generally place the upper limit of the habitable zone at around 1.7 au. Note that the habitable zone thus defined includes the orbit of the Earth (which is fortunate!) but also that of Mars: a planet more massive than Mars, with a thick atmosphere of carbon dioxide and water vapor, generating a sufficient greenhouse effect but regulated by the CO_2 cycle *via* carbonates and silicates, could *a priori* have a surface temperature compatible with the presence of liquid water. The fact that Mars is currently devoid of liquid water on its surface also illustrates the limits of the notion of habitable zone: it is not sufficient for a planet to be in this zone for liquid water to flow on its surface! Its atmospheric composition and surface pressure must be suitable for this.

Finally, let us not forget that the luminosity of the Sun increases with time. The habitable zone of the Solar System has therefore moved outward since its origin. Since the received solar flux is inversely proportional to the square of the heliocentric distance, we can estimate that 4 billion years ago, when the solar flux was 70% of its present value, the habitable zone was between 0.80 and 1.4 AU. Mars was then outside this zone, which illustrates the "paradox of the young Sun" that we mentioned above (see chapter 5). In the future, as solar brightness increases, the habitable zone will move away from the Sun and the Earth will in turn move out of it; it is estimated that within a billion years, the Earth will cease to be habitable because its surface temperature will be too high.

Figure 7.2 shows the distance of the habitable zone from the central star for the different classes of main-sequence (dwarf) stars. The Solar System is represented in front of type G2V which is that of the Sun. Stars A and F, which are more massive than the Sun, also have shorter lifetimes, less than a billion years; because they are brighter, their habitable zone is pushed to greater distances. Stars K and M, smaller and less massive than the Sun, are also the most numerous, and their lifetime is much longer. We will see later (chapter 8) that these stars are particularly interesting because their habitable zone is close to the star and the exoplanets that may be located there can be detected relatively easily when transiting in front of the star, because of their reduced period of revolution.

7.2   A Past Ocean on Venus?

Could the planet Venus have been habitable in the past? The question deserves to be asked. It is once again the luminosity of the "young Sun", 4 billion years ago, that is at issue. We have seen that it was, at the time, about 70% of its present value. At the level of the orbit of Venus, assuming for Venus an albedo comparable to that of the Earth today (0.3), the equilibrium temperature must have been close to 280 K (7°C), compatible with the presence of liquid water.

We have seen (chapter 4) that the atmosphere of Venus was rich in water some 3–4 billion years ago. The conditions could thus have been met so that an ocean of liquid water existed on Venus 3 or 4 billion years ago. What happened next? The solar flux gradually increased, causing the water to vaporize, then dissociate by solar ultraviolet radiation, and then the hydrogen atoms escaped. The absence of water may have prevented the development of plate tectonics as observed on Earth, where plate subduction uses hydrated materials. The absence of plate tectonics also prevents the recycling (through volcanism) of carbon dioxide trapped in the rocks and thus a regulation of the greenhouse effect as it occurs on Earth.

However, if an ocean of liquid water was present on the surface of Venus for several billion years, could life have appeared and developed there? There is nothing to exclude it *a priori*, but nothing can confirm this hypothesis. As we have seen above (chapter 3), all possible traces of life must have been erased with the remodeling of the surface due to volcanism (figure 7.3).

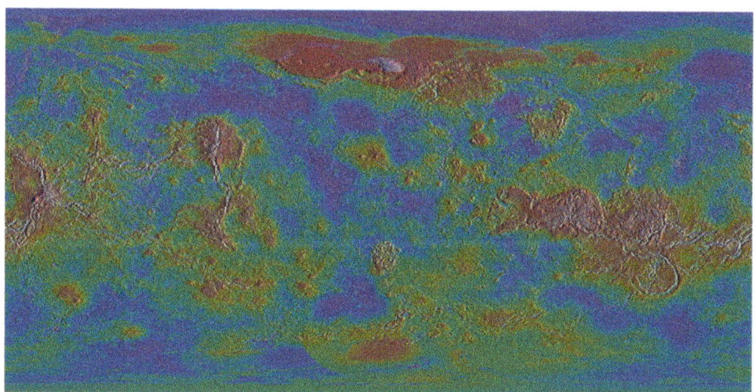

FIG. 7.3 – Mapping of the topography of Venus carried out by radar during the *Magellan* mission. The surface of Venus is entirely covered by volcanoes whose age is estimated at a few hundred million years, as shown by the absence of meteorite impact craters. Unlike Mars, there is therefore no hope of finding there any vestiges of the conditions that prevailed 3 billion years ago. © USGS Astrogeology Science Center.

However, some researchers at the University of Wisconsin are not giving up. Not being able to probe the surface, they turn to the clouds, where the conditions of temperature (230–300 K, or −47 to 27°C) and pressure (1–1.5 bar) are milder and would not be incompatible with microbial life. For decades, researchers have found that Venus clouds absorb solar ultraviolet radiation at a particular wavelength, but have not yet been able to unambiguously determine the nature of the absorbent.

Could it be bacteria? There is no evidence at present that this is possible, but it cannot be ruled out either. It should be noted, however, that there are terrestrial microorganisms surviving in environments dominated by carbon dioxide, rich in sulfuric acid and containing iron. It would be necessary to imagine, within the clouds of Venus, a metabolism based on iron and sulfur, in which the oxidation of these compounds would be coupled with the reduction of CO_2 in the absence of oxygen (figure 7.4). Life may have migrated from the surface to the clouds as the planet's temperature rose; the residence time of the microorganisms would have been sufficient to allow cell division before the cells sank and were destroyed. In 2020, a strong controversy arose in the scientific community, following the announcement of the tentative detection of phosphine PH_3 within the cloud deck. This detection would have been interpreted as a support to the scenario mentioned above. However, most scientists are very skeptical about this result.

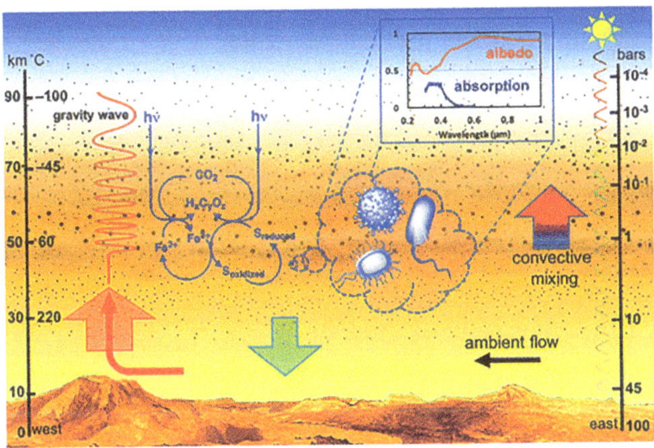

FIG. 7.4 – Diagram summarizing the ideas put forward in an article by S. Limaye *et al.* (2018) on the possibilities of microorganism-based life in the cloud layer of Venus. These microorganisms could survive by reduction of carbon dioxide from the oxidation of iron and sulfur compounds, or from iron and sulfur-based metabolism. They could remain suspended in the atmosphere, thanks to upward convective movements or gravity waves, for a sufficient time to ensure cell division, before falling to the surface. If present in clouds, the microorganisms could be the source of unexplained spectral signatures observed in the ultraviolet. Let us recall that this theory is for the moment only a hypothesis, devoid of any observational basis, whose validation will eventually only be possible by sending a probe capable of measuring *in-situ* the chemical composition of clouds. © S. Limaye *et al.*, *Astrobiology* 18, 1181 (2018).

This is a seductive suggestion, which deserves to be mentioned, even if today it is not based on any tangible facts. There is only one way to find out more: to resume space exploration of Venus, after *Venus Express* and *Akatsuki*, sending vehicles (balloons, drones...) capable of performing *in-situ* analysis of their chemical composition to the clouds. The technological challenges are much less than in the case of a descent probe that has to land on the surface, and the prospects of such a mission in terms of exobiology could be very significant.

7.3   Searching for Traces of Life on Mars

As we have seen (chapter 3), the search for life on Mars is not new. After the myth of the Martian canals, after the disappointment of the exploration carried out by the *Viking* mission, what about today? We will see that the quest for life on Mars is still alive and well.

The experiments conducted by *Viking* showed the absence of microorganisms, and even the absence of organic molecules, on the surface of Mars. There is a simple explanation for this: exposed to sunlight, organic matter would be immediately destroyed by the ultraviolet radiation that penetrates the atmosphere to the surface. The suspected oxidizing agent at the time, hydrogen peroxide, was finally identified some twenty years later. The search for life forms on the surface therefore appeared to be illusory.

However, as we have seen (chapter 5), there are multiple indications for the presence of liquid water in the past. These clues, which appeared as early as the *Viking* mission with the geological signatures of valley networks, were reinforced with the discovery, at the end of the 1980s, of deuterium enrichment from measurements made from Earth, and then with the resumption of space exploration at the end of the 1990s: possible existence of a boreal ocean (*Mars Global Surveyor*), discovery of permafrost reservoirs under the poles (*Mars Odyssey*), presence of clays in ancient terrains (*Mars Express*), and finally, evidenced by stratigraphic examination of the reliefs of the past, presence of subterranean water in various locations (figure 7.5). Recently, a team has announced, from radar data from the *MARSIS* instrument on board the *Mars Express* probe, the discovery of an aquifer, about 20 km wide, which would be located at a depth of one or two km below the southern polar cap of Mars. This result, if confirmed, could have important implications in terms of exobiology; however, the fact that the water is liquid at this shallow depth undoubtedly implies a very high salinity, *a priori* unfavorable to the existence of living microorganisms.

At the same time, the discovery of recurring stripes on the slopes of some craters (*recurring slope lineae*) has been interpreted as the possible presence of pockets of

FIG. 7.5 – The layered structure of the slopes of the Endurance Crater Burn Cliffs, observed by the American robot *Opportunity*, suggests that they were shaped by the presence of liquid water at the subsurface. From Grotzinger *et al.*, *Earth and Planetary Science Letters* 240, 11 (2005), © NASA.

salt water currently below the surface (figure 7.6), but this analysis remains controversial. More generally, the geological and stratigraphic record as a whole seems to show that, while water flowed in abundance until the end of the Noachian (3.5 billion years ago), it was much more episodic during the Hesperian and disappeared during the Amazonian, 3 billion years ago. So if ever life appeared on Mars, it is in the first billion years that we must look for its remains.

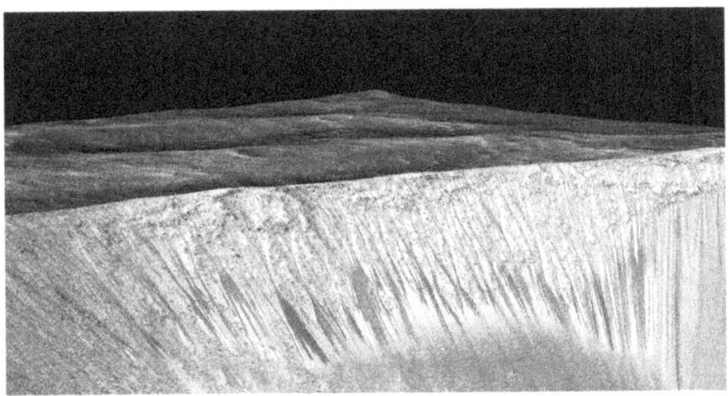

FIG. 7.6 – These parallel black streaks on the sides of the Garni Crater are called *Recurring Slope Lineae* (RSL). Identified on several dozen sites by the camera of the *Mars Reconnaissance Orbiter* mission, they could indicate the presence of liquid water below the surface. It would necessarily be very salty, a condition for the presence of liquid water to be compatible with the temperature there. However, according to some scientists, RSL could also be formed by the motion of sand grains in the absence of water. © NASA.

After the "Follow the water!" program set up by NASA at the beginning of this century to search for the signatures of liquid water, scientists have focused, in a second step, on determining the habitability criteria that would make an environment particularly conducive to the appearance of life. These criteria, based on what biochemistry teaches us, are the following: presence of liquid water, presence of nutrients (C, H, N, O, P, S), low acidity, low salinity, and presence of iron and sulfur in different oxidation states. It was the combination of these positive criteria that allowed the *Curiosity* mission scientists to conclude that the Yellowknife Bay site was a "habitable environment" some 3.5 billion years ago.

Beyond the habitability criteria, has the *Curiosity* rover found more tangible signs in favor of organic chemistry? For a long time, organic molecules were searched for without success. In 2015, chlorobenzene and dichloroalkanes were detected for the first time. But their abundance is low, and the organic matter that reacted with the Martian chlorine could be of external origin (comets, meteorites or interplanetary dust). Another important, but still controversial, discovery of *Curiosity* is the detection of a temporary emission of methane, at the level of 7 parts per billion in the atmosphere, during a few months. This jet, of unexplained origin, would be added to a continuous emission ten times weaker that seems to present seasonal variations. However, more recently, the *Trace Gas Orbiter* determined a very stringent upper limit for the amount of methane in the atmosphere above ten kilometers, significantly lower than the *Curiosity* detection, which refers to the surface level. These discoveries have rekindled a debate that has shaken the community for more than a decade, following a possible detection of methane on Mars. The presence of methane was quite unexpected, given the highly oxidizing nature of the Martian atmosphere, dominated by carbon dioxide; on Earth, methane is largely biogenic in origin. On Mars, its origin could be abiotic (*e.g.* by the outgassing of methane clathrates formed underground in the planet's past). The tentative discovery of Martian methane therefore has important implications in terms of exobiology, and the debate is not closed.

Let us return to the crucial question: if liquid water was abundant on the surface of Mars in the past, did it stay there long enough, and in conditions allowing the emergence of life? In the case of Yellowknife Bay, according to the authors of the research conducted by *Curiosity*, the length of time during which the conditions of habitability were met is unfortunately very uncertain (between 100 and 10 000 years). It is here that the "paradox of the young Sun" arises again. We have seen that, 3.5 billion years ago, the solar luminosity was lower than today, to the point that it was very difficult to imagine, at that time, an equilibrium temperature compatible with the presence of liquid water, therefore higher than 0°C. The generally preferred path – that of a generally cold and dry climate, interspersed with hotter and wetter periods, probably caused by volcanic episodes – is not *a priori* the most favorable scenario for the emergence of life, since longevity is one of the criteria retained for its appearance and maintenance. However, it can be noted that, as in the case of the Earth, a period of glaciation is not necessarily fatal if liquid water can continue to stay under the ice during cold periods, thus allowing the possible survival of potential microorganisms.

Let us suppose that microbial life developed at the beginning of the history of Mars, when water flowed in abundance. We have seen that life on Earth has a strong resilience. In the case of Mars, with the cooling of the planet and the disappearance of liquid water on the surface, could life have taken refuge below the surface, in environments protected from solar radiation, and adapted to hostile conditions? Lava tubes, formed on the surface of Mars during volcanic eruptions when a lava flow spreads and shrinks as it cools, could be such refuges. Oxidants such as hydrogen peroxide and perchlorate could in theory, in the absence of solar energy, provide the necessary energy for microorganisms, possible equivalents of the extremophiles that have been identified in the Siberian permafrost. For the moment, we have no element indicating their existence, but the question remains open and the exploration of Mars is more than ever on the agenda.

7.4   Other Niches in the Solar System

In all of the above, we have privileged the study of the terrestrial planets, which appear to be the most favorable environments for the emergence of life. There are, however, other potential niches in which the elements considered essential for the appearance of life can coexist: liquid water, nutrients (C, H, N, O, P, S) and the source of energy. This is the interior of some of the satellites surrounding the giant planets.

To better define the different types of habitats we may be confronted with, a new classification was introduced by the astronomer Helmut Lammer and colleagues in 2009 (figure 7.7).

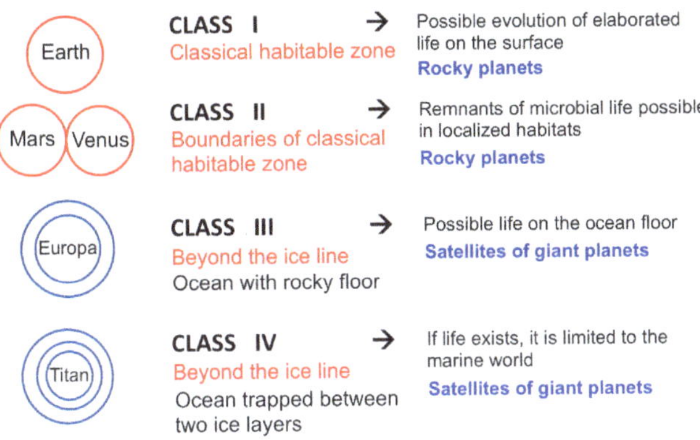

FIG. 7.7 – The different habitat classes, according to the classification of Lammer *et al.*, *Astronomy & Astrophysics Review* 17, 181–249 (2009).

Class I concerns environments in which complex multi-cellular life has developed; this is the case of the Earth and, to date, of it alone, but this class could someday extend to rocky exoplanets located in the habitable zone of stars close to the solar type (G-type stars or K or F stars whose mass is close to that of the Sun), where habitability conditions could be maintained over the long term.

Class II concerns sites where life could appear, but under conditions of stellar radiation or excessive radiation likely to lead to its disappearance. This is the case of Venus and Mars (chapter 4); it could also concern exoplanets located in the habitable zone of dwarf stars of type K and especially M, which very often emit X-rays.

Class III includes habitats corresponding to icy bodies (outer satellites, trans-Neptunian objects or, perhaps, satellites of giant exoplanets) whose inner liquid ocean, containing reducing elements such as hydrogen, would be in direct contact with the silicate soil, allowing an exchange of energy and the possible development of a complex chemistry. We will see that this is the case for the outer satellites Europa and Enceladus.

Finally, class IV characterizes sites where liquid water could be present, but confined between two layers of ice; this is probably the case of several satellites: Ganymede, Callisto and perhaps even Titan. It is therefore the case in which the appearance of life appears, *a priori*, the least plausible.

How were the outer satellites of the Solar System formed? We have seen (chapter 2) that the giant planets of the Solar System were formed far from the Sun, by accumulation of a core essentially composed of ice and heavy elements. This nucleus proved to be massive enough that its gravity could capture the surrounding gas, essentially hydrogen and helium. When this gas collapsed, a disk formed around the equatorial plane of the planet, like the protosolar disk around the young Sun. Within this disk, satellites were born, forming miniature solar systems. Given the abundance of hydrogen and oxygen atoms in the Universe, water ice is a preponderant element of these satellites, especially in the case of Jupiter and Saturn. The largest of them have a differentiated internal structure, with a mantle of ice and a central core made of silicates and metals. For some of them, the gravity field of their planet is sufficient to produce variable tidal forces that result in an increase in their internal temperature, up to a level sufficient for water to be in liquid form. In the most favorable cases, according to internal structure models, the liquid ocean, topped by an ice layer, could be in direct contact with the silicate soil: this is probably the case of Europa, satellite of Jupiter, and Enceladus, satellite of Saturn.

Europa is the second of the Galilean satellites, in terms of distance from Jupiter. The closest one, Io, is the seat of intense active volcanism, due to the violent tidal effects that have the effect of generating a volcanism that constantly renews its surface. Europa, like the other two Galilean satellites Ganymede and Callisto, is covered by a layer of ice. All were first explored by the *Voyager 1* probe in 1979 during its flyby of Jupiter. Astronomers were immediately intrigued by the striated appearance of the surface of Europa, which suggested the existence of moving plates above a viscous, perhaps even liquid medium; the *Galileo* mission, in orbit around Jupiter nearly twenty years later, confirmed this analysis and, moreover, discovered the existence of an induced magnetic field, probably generated within an ocean of salt water and therefore conductive of electricity. Moreover, according to internal structure models, the liquid ocean could be in direct contact with the silicate soil, which would theoretically make possible energy exchanges and the development of a prebiotic or even biotic chemistry (figure 7.8). As for the two other Galilean satellites further away from Jupiter, Ganymede and Callisto, they too probably have an internal liquid ocean, but, according to theoretical models, it is probably trapped between two layers of ice, which reduces the perspectives from the point of view of exobiology. The interest of Europa's internal ocean alone is sufficient to have justified the continuation of its space exploration: two missions for the flybys of Europa and the exploration of Galilean satellites are being prepared on both sides of the Atlantic, at ESA (*JUICE*) and NASA (*Europa Clipper*) for launch in the 2020s.

Even further from the Sun, in Saturn's system, the surprise came from Titan, its largest satellite. Before the space age, the existence of its atmosphere, unique in the Solar System, had already been demonstrated, but its composition and surface pressure were unknown. In 1980, the *Voyager 1* probe revealed that this atmosphere, rich in molecular nitrogen and with a surface pressure comparable to that of the Earth, presented curious analogies with our own planet. What's more, it discovered a number of complex molecules, hydrocarbons and nitriles, known as "prebiotics", those involved in the reactions leading to the formation of amino acids and the building blocks of life. Could Titan be a laboratory for prebiotic chemistry? It was this question among others that justified the ambitious *Cassini-Huygens* mission, operated jointly by NASA and ESA in 1997, in operation around Saturn and Titan from 2005 to 2017. Its success has been immense, both technologically and scientifically.

Titan's atmosphere has been studied in depth for nearly two seasons. The methane cycle, which is present in its three phases, was highlighted, with its emission from the interior in the form of cryovolcanism, its condensation in the form of

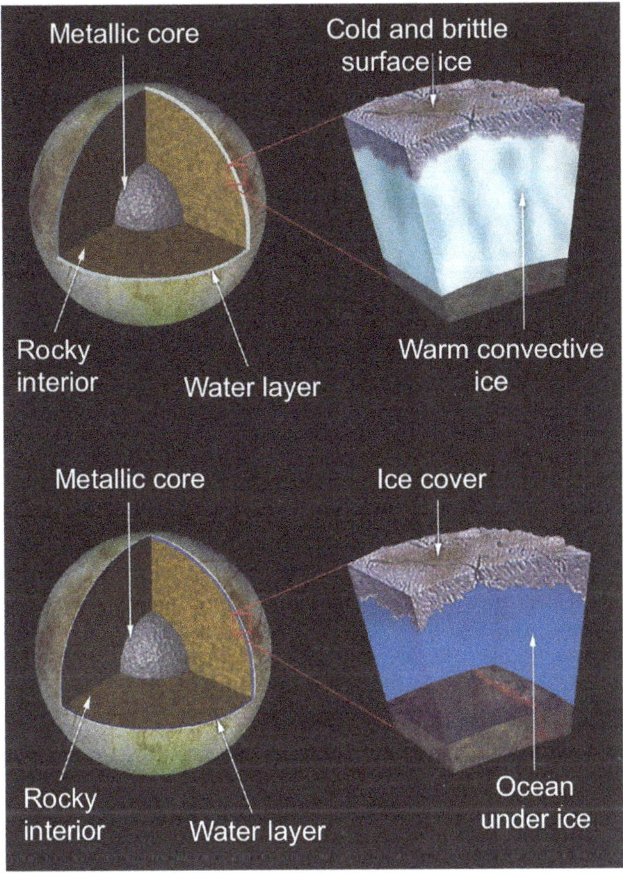

FIG. 7.8 – Two possible models of the internal structure of Europa, compatible with *Galileo* probe data. The superficial ice layer overlies (top) a thick layer of warm convective ice or (bottom) an ocean of salt liquid water. In both cases, the water would be in contact, at depth, with the silicate core. Adapted from A. Coustenis and T. Encrenaz, *Life beyond Earth*, Cambridge, 2013.

clouds, its photodissociation into various hydrocarbons and its seasonal storage at the surface in the form of lakes at high latitudes (figure 7.9). *Cassini*'s observations have also shown the presence of a liquid ocean below the surface of Titan, but, as in the case of Ganymede and Callisto, it is probably trapped between two layers of ice.

Among the many results of the *Cassini* mission is an unexpected discovery: that of a liquid ocean beneath the surface of Enceladus, a small icy satellite of Saturn. In 2005, the *Cassini* probe detected gas ejections near the south pole of the satellite (figure 7.10). Their analysis by the instruments of the probe revealed not only water vapor but also organic compounds, molecular nitrogen and carbon dioxide; in 2016, molecular hydrogen was also detected. The proposed origin of this cryovolcanism is a geyser from a pressurized liquid ocean, which would extend to the silicate nucleus of

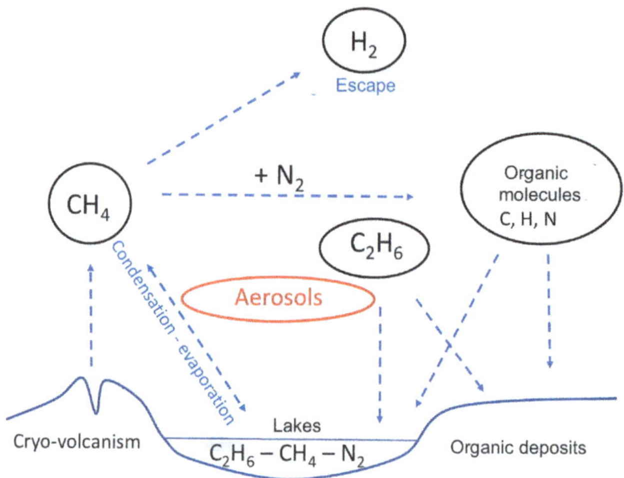

FIG. 7.9 – The methane cycle on Titan. Methane is emitted from the interior of the satellite by cryovolcanism and evaporates from the lakes present in winter at high latitudes. In the atmosphere, it is dissociated into various hydrocarbons and combines with nitrogen atoms to form aerosols that condense and fall back to the surface and into the lakes. According to F. Raulin, LISA and A. Coustenis and T. Encrenaz, *Life beyond Earth*, Cambridge, 2013.

FIG. 7.10 – This backlit image of Enceladus taken in November 2005 by the *Cassini* probe camera shows jets emitted over the South Pole. This region, whose temperature is slightly higher than its environment, is streaked with fractures through which these jets escape, mainly made up of water vapor partially condensed into ice. The solid part of these emissions feed the E-ring of Saturn, whose orbit coincides with that of Enceladus. © NASA.

the satellite. The internal energy would come from the tidal effects generated by the presence of Saturn and other satellites, notably Dione, but the models have difficulty explaining the rise in temperature necessary for water, even mixed with other constituents such as ammonia, to be in liquid form. According to another hypothesis, the energy would come from the turbulence generated by the oscillation of the ice crust over the liquid ocean; the question remains open. From an exobiology point of view, Enceladus could prove to be an even more fascinating niche than Europa, because its ocean could be closer to the surface than in the case of Europa. However, Enceladus is also twice as far from the Sun as Europa and no new space mission is currently planned to continue its exploration.

The list of potential niches associated with the inner oceans of ice bodies does not stop at the objects we have mentioned above. Other external satellites are likely to contain a liquid ocean. Other surprises may still await us. For example, in 2014, the small planet Ceres was the object of an unexpected discovery: the European space observatory *Herschel* discovered an emission of jets of water vapor and ice at the surface (figure 7.11). The following year, the U.S. space mission *Dawn* discovered water ice and hydrated materials (including carbonates and clay) on the surface, suggesting the possible existence of a reservoir of liquid water beneath the ice crust. The asteroid is indeed massive enough to have a differentiated internal structure, with a central core surrounded by a reservoir of liquid water or viscous ice, topped by a layer of ice. Ceres could be an ancient trans-Neptunian object, formed beyond Neptune's orbit, which would have been ejected towards the interior of the Solar

FIG. 7.11 – The Ceres asteroid seen by the camera of the *Dawn* probe. The white spots correspond to salt deposits in the Occator crater. They are probably the trace of ancient geysers. © NASA.

System in the Main Asteroid Belt during the great planetary migrations (see chapter 2). Ceres could thus also be a potential niche for exobiology, and the list of such potential habitats is probably far from being closed (figure 7.11).

Chapter 8

How to Search for Life on Rocky Exoplanets?

Our search for extraterrestrial life within the limits of the Solar System has not been very fruitful: whether we are looking at Mars or external satellites, we can only hope to find, in the best case, embryonic traces of life. In the case of Mars, research is focused on possible fossil forms of life that might have appeared some four billion years ago. On the other hand, inside the outer satellites that may harbor a liquid ocean beneath their icy surface, such as Europa around Jupiter and Enceladus around Saturn, embryonic life could be sought for, like those found near hydrothermal springs at the bottom of the Earth's oceans. No hope therefore to detect any evolved form of life in the Solar System.

It is here that the discovery of extrasolar planets, over the last twenty-five years, has completely renewed this field of research. We know today that, within our Galaxy, we are surrounded by billions of rocky exoplanets, whose physical and orbital conditions are infinitely diverse; among this multitude, a non-negligible fraction probably lie in the habitable zone that we mentioned in the previous chapter, the zone in which the temperature is compatible with the presence of liquid water. The consequences for the search for extraterrestrial life are immense, and the field of possibilities opens up to infinity.

8.1 The Discovery of Exoplanets: Where Do We Stand?

Before exploring these new possibilities in more detail, we must first look back at what has been a real revolution for astronomers over the last two decades: the discovery of a very large number of planets around other stars. Let us start with a first assessment: currently, at the beginning of 2021, our catalogs contain about 4700 identified and confirmed exoplanets, to which are added several thousand candidates. It is estimated that, on average, each star in our Galaxy is surrounded by an exoplanet. Planetary systems, including at least two or more exoplanets, are also

DOI: 10.1051/978-2-7598-2563-9.c008

numerous: there are currently more than 770 of them, and the number of planets surrounding a star can be as high as seven!

Another striking and unexpected result: the extraordinary diversity of these planets. A major surprise appeared as soon as the first exoplanet around a solar-type star was discovered in 1995: there are giant exoplanets in the immediate vicinity of their host star (figure 8.1)! Let us recall that this situation is completely different from that of the Solar System: in this case, as we have seen, giant planets formed far from the Sun, beyond the "ice line" which corresponds to the condensation temperature of water. This peculiarity is explained by the mode of formation of planets by accretion around a solid core (see chapter 2). To explain the new discoveries, the theorists used a new scenario, the migration that moves giant exoplanets initially formed far from their star inward. This scenario was reinforced by the discovery, thanks to observations with the ALMA millimeter telescope array, of young exoplanets migrating within a protoplanetary disc. In parallel with these discoveries, researchers have shown that migration has also played an important role within the Solar System itself. However, this migration was less radical than in the case of many exoplanets: the planet Jupiter, after moving closer to the Sun down to Mars' orbit, then moved outward under the interaction with Saturn's attraction, thus preserving the planets of the inner Solar System and in particular the Earth (see chapter 2).

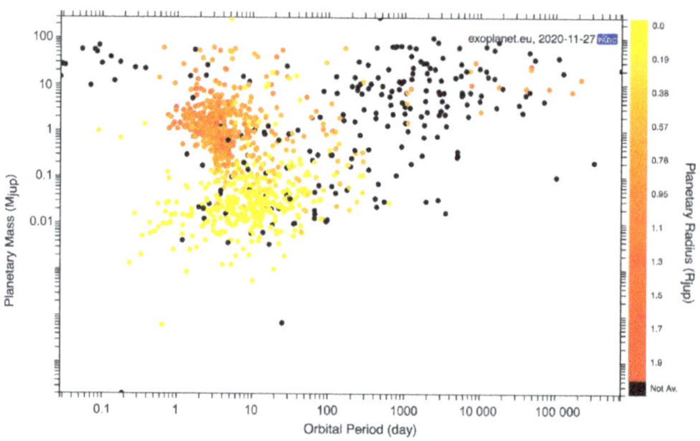

FIG. 8.1 – Distribution of the mass of confirmed exoplanets as a function of their period of revolution. We see that there are a significant number of giant exoplanets with very short periods in the immediate vicinity of their host star; these are the easiest to detect and were discovered first. © europlanets.eu.

The diversity of exoplanets concerns both their physical and orbital characteristics. Some exoplanets (figure 8.2), close to their star, have a very low density (less than 0.1 g/cm^3, or a tenth of the density of water): they are giant exoplanets called "inflated", strongly irradiated; others, very small on the contrary, have a very high density (up to 50 g/cm^3, or ten times the Earth's density): these could be residues of giant planets that have migrated towards their star, having lost all their gas following irradiation by the neighboring star and reduced to their central core, made

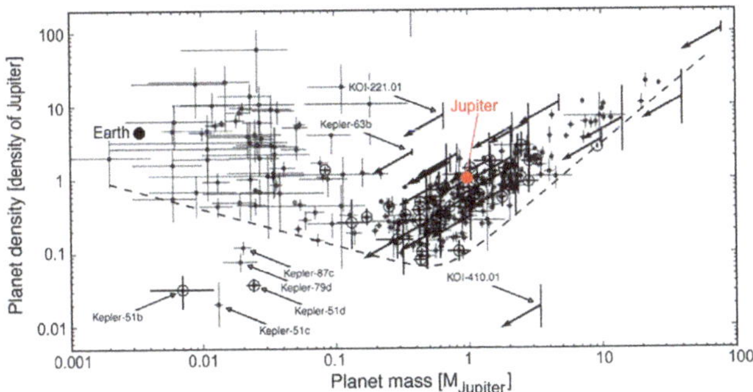

FIG. 8.2 – Distribution of the density of exoplanets as a function of their mass. The arrows indicate the objects for which only a lower limit of mass is known. We see giant "inflated" exoplanets (center) and very small and very dense objects (top left). Adapted from Santerne *et al.*, *Astronomy & Astrophysics* 587, A64 (2016).

very dense by the effect of the compression that took place when the giant planet was very massive.

Another surprise: exoplanets do not fall into two quite distinct categories, the "rocky" and the "giant", as in the case of the planets of the Solar System: we observe a continuity of objects from masses and radii smaller than those of the Earth to those of Jupiter and beyond, small planets being largely more numerous than the giant ones. New classes appear, the "super-Earths", presumed to have a rocky surface, with a mass of up to ten Earth masses and a radius twice as large, and the larger "exo-Neptunes", presumed to have a hydrogen atmosphere (figure 8.3). It should be pointed out that density alone is not sufficient to determine whether an exoplanet is rocky or gaseous; to answer this question, the composition of the atmosphere must be known; we shall return to this point later.

Let us now talk about orbits; here again, we are far from the regularity of the Solar System whose eight planets wisely orbit in close proximity to the plane of the ecliptic (defined by the Earth's orbit), on quasi-circular trajectories around the Sun. Some exoplanets have circular orbits, but others have very high eccentricities, up to 0.9 (figure 8.4). As for multiple systems, they also sometimes contain planets located on very inclined or eccentric orbits. However, among the 770 multiple systems detected, we have not yet found a system resembling our Solar System, which is thus far from representing a widespread model in the Galaxy.

Let us now move on to stars other than the Sun. Within the Galaxy, binary stars are frequent, more frequent than isolated stars like the Sun. Exoplanets associated with binary stars have also been discovered, either orbiting outside the system or inside around one of the stars in the pair. The diversity of configurations is thus immense, suggesting that several planetary formation scenarios could be at work: accretion around a nucleus as in the case of the Solar System, but also gravitational instability within a cloud for the largest planets, as is the case for star formation. The obvious conclusion is that the phenomenon of planetary formation is clearly widespread in the Universe.

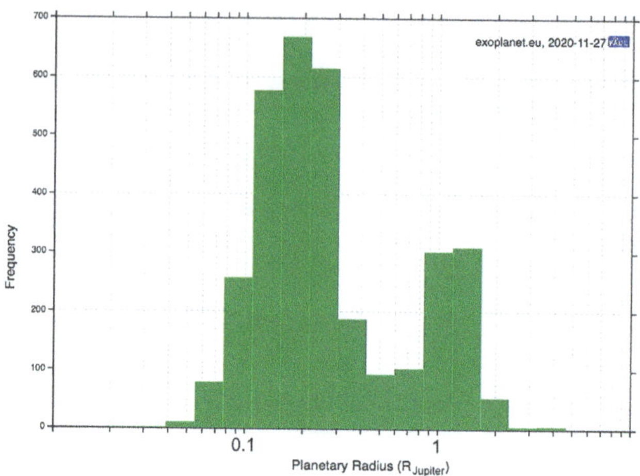

FIG. 8.3 – Distribution of exoplanets as a function of their radius. We see that there are a large number of objects with a radius intermediate between that of the Earth (*i.e.* 0.09 R_Jupiter) and a few tens of terrestrial radii (*i.e.* 0.3–0.4 R_Jupiter): these are the "super-Earths" and the "exo-Neptunes". The Jovian exoplanets form a very distinct minority population. © exoplanets.eu.

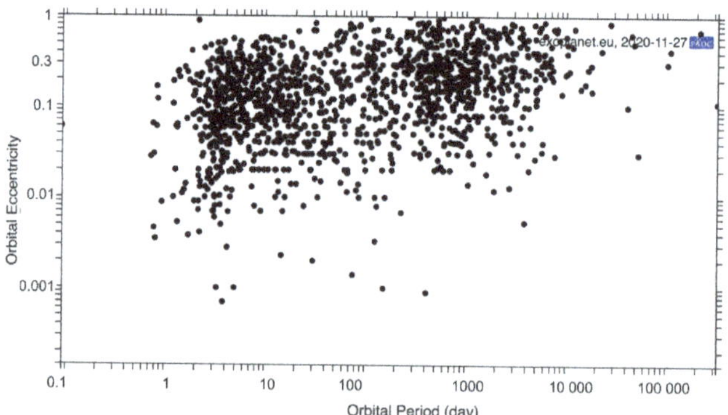

FIG. 8.4 – Distribution of the eccentricities of exoplanets according to their period of revolution. A wide range of eccentricities can be observed, whatever the distance to the star. © exoplanets.eu.

Finally, to complete this bestiary, we must mention the discovery of a particular category of exoplanets: the so-called "floating" exoplanets that are not located near a star. How did they get there? They could be objects formed within the interstellar medium by gravitational collapse like stars, but not massive enough to reach the brown dwarf stage; or else, these planets could have been ejected from their initial star system following gravitational interactions due to other exoplanets of this system or to the proximity of a neighboring star. These are, summarized in a few words, the main lines of what we know today about extrasolar planets. The large number of exoplanets already discovered and the diversity of their physical and orbital characteristics encourage us to further analyze their possible habitability. But before opening this section, we must briefly describe how these results were acquired: the history of the first discoveries and the different detection methods used, with their advantages and limitations.

8.2   The Exoplanet Concept: An Old Idea

As we have seen above (chapter 2), the idea of the possible existence of planets around other stars is not new. Traces of it can be found as early as Antiquity, in particular in a writing by the philosopher Epicurus (ca. 342–270 B.C.) and later, in the wake of the Copernican revolution, with Giordano Bruno (1571–1600) in particular. Following him, many philosophers and astronomers speculated on the possible existence of extraterrestrial life, within the Solar System or beyond, from Johannes Kepler (1571–1630) to Camille Flammarion (1842–1925) in the last century.

For all these authors, it was still only speculation: there was no question of attempting a direct observation of a planet in orbit around a star. Indeed, let us take the case of the Sun and Jupiter, the largest planet in our Solar System. The diameter of Jupiter is ten times smaller than that of the Sun, and the planet is about a thousand times less massive. In the visible range, the radiation it emits is the radiation reflected by the Sun; seen from outside the Solar System, it is more than a million times less bright than the Sun; moreover, seen from a nearby star a few light years away, the angular distance between the Sun and Jupiter is only a few tenths of an arc second. Impossible to distinguish it from its star, it is too close to it and is drowned in starlight.

The astronomers of the last centuries understood this problem well and, in order to get around it, considered using indirect methods. The first one consists in observing the motions on the sky of the star-planet system (figure 8.5): the star

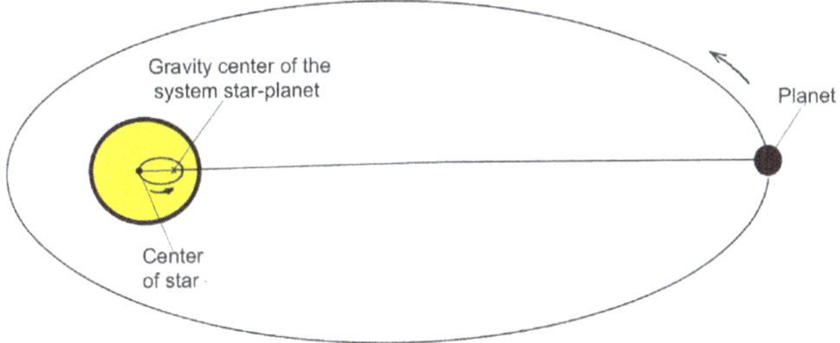

FIG. 8.5 – Diagram representing the system consisting of a star and a planet orbiting around it. The center of gravity of the system is slightly shifted with respect to the center of the star, which follows an ellipse of small amplitude around this point. It is this small motion that astronomers have tried to measure, first (unsuccessfully) by astrometry and then (successfully) by velocimetry.

orbits around the center of gravity of the system with a period that is the period of the planet around the star. The method has been successfully used since the beginning of the 19th century in the case of double star systems; the principle here is the same, with the difference that the orbit described by the star is much smaller than that described by the planet, whose mass is much lower than that of the star. It is therefore necessary to measure a very small motion, during a period of time *a priori* unknown, by locating it on the sky with respect to the neighboring stars.

Thus, in the course of the 20th century, several detections of exoplanets were successively announced and then questioned. The most famous case is that of Barnard's star, observed by Peter van de Kamp (1901–1995) for several decades. His discovery of two planets around this star was finally invalidated: the announced effect was due to an instrumental error related to the telescope he was using. At the end of the 20th century, astronomers realized that the measurement of the motion of a star in the sky (this is called astrometry), could not be sufficiently precise given the means of the time to allow the detection of an exoplanet. They then turned to another method: to measure the periodic motion of the star relative to the center of gravity of the star-planet system, they used the variations of its velocity relative to the observer. This is called velocimetry. Starting in 1995, this method allowed the discovery of hundreds of exoplanets.

8.3 Early Discoveries

Let us place ourselves at the beginning of the 1990s. There was still no known exoplanet, but there was renewed interest in this research. Indeed, for about ten years now, astronomers, thanks to the *IRAS* infrared satellite in 1983, had discovered the existence in large numbers of protoplanetary disks around young stars. Like the scenario of the formation of the Solar System (see chapter 2), the formation of a disk seemed to be a common step in the star formation scenario. Therefore, the formation of exoplanets within this disk appeared as the next step, plausible or even probable.

Curiously, the first discovery of exoplanets did not concern planets around young stars, but around a very particular type of star at the end of its life, a pulsar ("pulsating star"). Pulsars are extremely dense stars, at the final stage of their evolution, which have the particularity of emitting in rotating a radio signal of very short period (which can be as short as a millisecond) and extremely stable. If the pulsar is surrounded by one or even several planets, variations in the pulse time arrival can reveal them. Thus, in 1992, the Polish astronomer Alexander Wolszczan announced the discovery of two exoplanets, each with a few Earth masses, around the "millisecond" pulsar PSR B1257+12. The announcement caused a sensation. On the one hand, it was the first discovery of an extrasolar planet; on the other hand, nobody had imagined the presence of planets around a pulsar, which is a neutron star resulting from the explosion of a supernova. This discovery proved that the formation of accretion disks was more frequent than expected, but also still poorly understood. We know today about forty exoplanets in orbit around pulsars. Considering the violent conditions that accompanied their formation and the abundance of X-rays emitted by the pulsar, they should hardly resemble the planets we know, and have very little chance of harboring life. Their discovery was nevertheless an important step in the search for exoplanets.

8.4   The Successes of Velocimetry

At that time, several teams have specialized in the search for low mass stellar objects. Their method is velocimetry: it consists in measuring with a very high degree of accuracy the radial velocity of a star with respect to the observer. The method rests on the Doppler–Fizeau effect, widely used in astronomy, which uses the displacement of the frequency (or wavelength) of the radiation emitted by a star as a function of its displacement relative to the observer. If the radial velocity of the star shows a periodic variation, this is the signature of the presence of a less massive companion, not detected in visible light. Note that this method is perfectly complementary to astrometry: if the planet's orbit is located in the plane of the sky, perpendicular to the viewing direction, the motion of the star is a circle or an ellipse on the sky that can be detected by astrometry, while velocimetry does not detect any change in speed; on the other hand, if the viewing direction is located in the plane of the orbit, the effect measured by velocimetry is maximum and astrometry detects the motion of the star as a segment on the sky. Both methods make it possible to determine a lower limit of the mass of the planet (because we do not know the angle under which the system is seen by the observer) as well as the period of revolution of the planet. In the search for the exoplanets, it was initially considered that both astrometry and velocimetry would require long-term observation programs, since it is necessary to cover a whole period of revolution of the exoplanet to confirm definitively its existence. As only giant planets were detectable with the observational means of the end of the 20th century, and since it was thought that these giant planets were always at large distances from their star with large revolution periods, it was expected that detection would require several years of observation; as an example, the period of revolution of Jupiter around the Sun is about twelve years.

Here a happy surprise awaited the astronomers. If the giant exoplanets in extrasolar systems had had the characteristics of those in the Solar System, it would have taken years of observation to detect them. But this is not the case: there are a large number of giant exoplanets very close to their star. As they are the easiest to identify, they were the first to be detected by velocimetry. In 1995, Michel Mayor and Didier Queloz, from the Geneva Observatory, announced the discovery of the first exoplanet orbiting around a solar-type star. The discovery was made using a high-precision spectrograph built by André Baranne, from the Marseilles observatory; operating in the visible range, it was attached to the 193 cm telescope of the Haute-Provence observatory (figure 8.6). This planet is 51 Peg b, a planet whose mass is at least half that of Jupiter, which revolves around its star (51 Peg a) on a circular orbit with a period of four days, at a distance of 0.05 au from this star. The news was like a bomb in the scientific community. Not only astronomers have

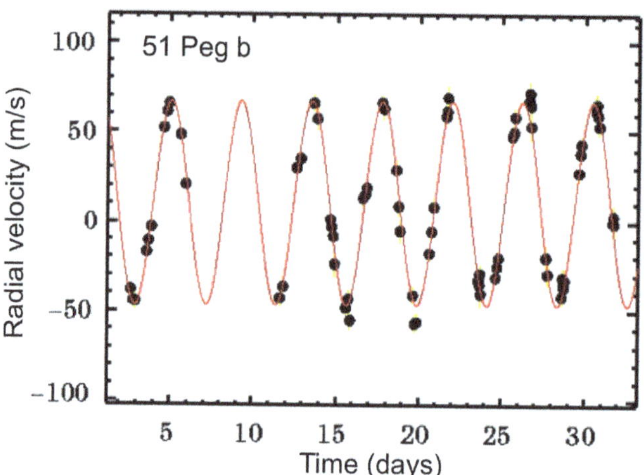

FIG. 8.6 – First discovery of an exoplanet around a solar-type star (51 Pegasi). The figure shows the periodic oscillation of the radial velocity of the star with respect to the observer, measured as a function of time. The period of the oscillation gives the period of revolution of the star and the planet around their center of mass, its amplitude provides a lower limit of the mass of the planet. In the case of 51 Peg b, this mass is at least half that of Jupiter and the period of revolution is 4 days, which corresponds to an orbital distance of 0.05 au. The discovery was made at the 193 cm telescope of the Observatoire de Haute-Provence, by M. Mayor and D. Queloz, from Geneva Observatory, and won them the 2019 Nobel Prize in Physics. Adapted from M. Mayor and D. Queloz, *Nature* 378, 355 (1995).

discovered a planet around a solar-type star, but this giant exoplanet very close to its star looked nothing like the planets of the Solar System. It was thus proven that if it is not unique, its configuration is not universal, which strongly questioned our understanding of stellar and planetary formation scenarios. In the following weeks, the detection announcements multiplied, involving teams from the United States and Canada. The planets so discovered were all giant exoplanets very close to their star; their existence was therefore not exceptional. That they are the first exoplanets detected is not a surprise: massive exoplanets with a short period of revolution are the easiest to detect. At the end of the 20th century, about a hundred exoplanets had been discovered, almost all of them by the velocimetry method.

8.5 A New Step: The Transit Method

The dawn of the 21st century marks a turning point in the exploration of exoplanets: a new detection method came into play, the transit method. The principle is the following: if a planet passes in front of its star, it creates a small obstruction on the stellar disc, and its light is diminished by a small amount (figure 8.7). In the case of Jupiter, whose diameter is one tenth of that of the Sun, the decrease in sunlight, seen from a neighboring star, is 1%. In the case of the Earth whose diameter is about ten

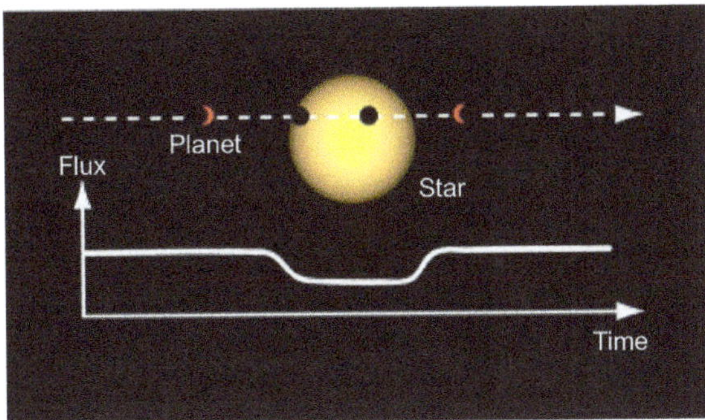

FIG. 8.7 – Diagram representing a planetary transit, *i.e.* the passage of the planet in front of its host star. The planet is invisible to the observer, but if we measure very precisely the light of the star as a function of time, we observe a slight decrease in the brightness of the star when it is obscured by the planet.

times smaller than that of Jupiter, the decrease in sunlight is only 0.01%. To detect the presence of an exoplanet in transit in front of its star, it is necessary to measure the light of the star with great precision (this is called photometry) during the longest possible time intervals. Current photometers installed on ground-based telescopes allow to measure stellar light with an accuracy of 1%; the detection of giant exoplanets is thus possible from the Earth. On the other hand, that of terrestrial type exoplanets (that of rocky exoplanets, which are called, according to their mass, "exo-Earths" or "super-Earths") generally requires measurements from space, where greater stability can be achieved. It is, however, possible to detect from the ground terrestrial type planets in transit in front of lowmass stars, which are very faint. Note, however, that the measurement of transits is not sufficient to definitively establish the existence of an exoplanet: in some cases, variations in the stellar light may be due to pulsations of the star itself. When the presence of an exoplanet in transit is suspected, velocimetry measurements must be made to confirm the detection. This yields a measure of the mass of the planet; added to that of its radius, obtained by the observation of the transit, one has thus access to the density of the object.

The year 1999 marks the first discovery of an exoplanet by transit (figure 8.8). It is, again, a giant exoplanet very close to its star, which bears the barbarian name of HD209458 b. The star itself is close to us and bright, which made observation easier. Later, the observation was repeated with the Hubble Space Telescope, which gave a better accuracy. From then on, observation campaigns from the ground multiplied, using medium size telescopes or arrays of small telescopes. The number of detections made from the Earth reached about 30 at the end of 2009; ten years later, it is several hundred.

The implementation of the transit method with terrestrial telescopes thus marks a new stage in the exploration of giant exoplanets. But, at the beginning of our century, it remained to discover the less massive exoplanets, those which are probably rocky and which interest us primarily for their potential exobiological

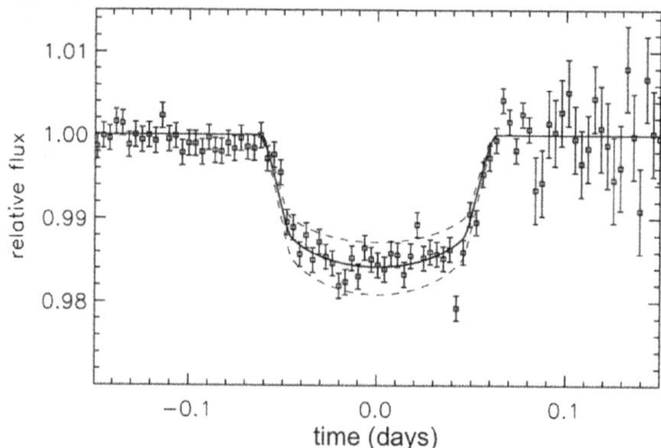

FIG. 8.8 – First discovery of an exoplanet by the transit method. The figure represents the light curve of the star HD209458 during the transit of the exoplanet HD209458 b in orbit around the star. The observations were made with the 10-cm diameter Schmidt STARE telescope located at High Altitude Observatory, Mauna Loa, Hawaii. From D. Charbonneau *et al.*, *Astrophysical Journal* 529, L45 (2000).

interest. One thus envisaged to place on a satellite a photometer capable of obtaining a measurement precision of the order of 10^{-4} over a long period of time. As it was now known that many exoplanets have a period of revolution of only a few days, a duration of observation of the order of one month was considered sufficient.

The first space mission dedicated to this program was developed and operated by the French space agency, CNES. *CoRoT* (COnvection, ROtation and Planetary Transits, figure 8.9) has a double scientific objective: the study of stellar oscillations and the search for exoplanets by transit. Both required the same instrumentation: an extremely precise and time-stable photometer. Launched in 2006 and in operation until 2012, it detected thirty exoplanets, among which the small rocky exoplanet CoRoT-7 b, which rotates around its star in less than a day, and the "cold" giant exoplanet CoRoT-9 b, whose period is 95 days. This pioneering mission already revealed the extraordinary diversity of the exoplanetary zoo.

Then came *Kepler* (figure 8.10), a mission launched by NASA in 2009. It operated nominally until 2013, then until 2018 (mission *K2*) with reduced capabilities following the loss of two reaction wheels. More powerful than *CoRoT*, the *Kepler* mission allowed a real leap forward in our knowledge of exoplanets. As in the case of velocimetry measurements, astronomers were served by chance. The presence of many exoplanets very close to their star favors the transit method for two reasons: on the one hand, the transit frequency of an exoplanet in front of its star is greater than in the case of the Solar System; on the other hand, the probability of such a transit is higher. At the end of the mission, *Kepler* had discovered more than 2600 confirmed exoplanets. This large data base made it possible to define statistics on the nature of the exoplanets: half of them ("super-Earths" or "Neptunes") have masses between 3 and 30 terrestrial masses; giant exoplanets represent approximately 30% of the total. The percentage of exoplanets of small mass is certainly very largely underestimated, these being the most difficult to detect.

FIG. 8.9 – Artist's view of the *CoRoT* satellite. © CNES.

FIG. 8.10 – Artist's view of the *Kepler* satellite. © NASA.

We have described above the two methods of exoplanet detection (velocimetry and transit) that have proven to be the most productive. Let us briefly mention other techniques that have also enabled detections: the observation of "gravitational transits" (the passage of a star with a planet in front of a distant source from which it deflects its rays by relativistic effect), which has enabled the detection of about 140 objects; direct imaging of giant exoplanets (generally young and relatively distant from their star), in full development, which has led to the identification of about 140 other objects. Finally, direct astrometry, whose performance was too limited in the case of ground-based observations, will enable the detection of a multitude of exoplanets, following the exploitation of star catalogs from the European *Gaia* mission launched in 2013, which measures the positions and velocities of more than a billion stars. Already, *Gaia* has shown that more than 30% of the nearby stars, and perhaps 50%, have an unseen massive companion, a planet or a brown dwarf star.

8.6   How to Search for Life on an Exoplanet?

Let us take up again the conditions we mentioned in the case of the quest for extraterrestrial life within the Solar System. We have identified three *a priori* necessary conditions: the presence of liquid water, carbonaceous matter and a source of energy. Let us project these conditions in the context of exoplanets. The easiest energy source to imagine is that of the radiation of the star around which the exoplanet orbits. The simultaneous presence of liquid water and carbonaceous matter implies the presence of a surface on the planet (giant planets like Jupiter have none). The existence of liquid water requires a temperature higher than 0°C. The upper limit is fuzzier because it depends on the pressure of the medium. In the case of the Earth, it is 100°C, but it can be a little higher if the pressure is higher, while remaining necessarily lower than that of the critical point of water, *i.e.* 374°C. In the case of the Solar System, we have defined the region in which water can be in a liquid state as the habitability zone; this notion can be extended to other types of stars (see chapter 7). We thus consider as priority candidates the rocky exoplanets belonging to a stellar system and located in the habitability zone of the host star.

How can we know if an exoplanet is rocky? The best index is that of its density, obtained from the measurement of its radius and its mass; our sampling will thus preferentially target exoplanets observed by transit. We have seen above (chapter 2) that, in the scenario of planetary formation by accretion, there is a limit mass, around 10 Earth masses, beyond which the gravity field of the planet is sufficient to capture, within the protoplanetary disc, the surrounding gas; the planet then becomes a giant. We will thus keep among our candidates the objects of mass lower than 10 terrestrial masses and having a density at least equal to 3 g/cm^3, compatible with the presence of carbonaceous, oxidized or silicate rocks.

Knowing the mass and the spectral type of the star, we can calculate the radiation it emits and therefore the distance at which the temperature will allow the presence of liquid water, thus knowing which of the exoplanets identified around this star are in the habitability zone. As we observed previously (chapter 7), the habitability zone is a function of the mass of the star: the more massive the star is, the further away the habitability zone is; the less massive it is, the closer the zone gets to it. In terms of habitability, the latter case is the more favorable for two reasons. On the one hand, with about 90% of the total stellar population, stars less massive than the Sun are the most numerous, which multiplies the probability of detecting exoplanets around these stars; moreover, they have a very long lifetime, longer than the age of the Universe. The period of revolution of their exoplanets is shorter, which makes observations by transit easier. However, there are disadvantages: some dwarf stars (especially red dwarfs) show eruptive phases with X-ray emission, which could be unfavorable to the development of life. In addition, when they are close to their star (with a period of less than a hundred days), rocky exoplanets are generally in synchronous rotation, *i.e.* always present the same face to the star, like the Moon to the Earth. This configuration, which implies marked and permanent contrasts between the day and night sides, must have strong implications in terms of exobiology.

That being said, do we know of any exoplanets that qualify? The answer is yes, there are currently about 50. The first announcement, dating from 2007, concerned two exoplanets located around the star Gliese 581, a dwarf star of type M, but this result was then reversed. In 2017, the *Kepler* mission identified about twenty of them. More recently, surveys from Earth have identified new candidates, including the famous TRAPPIST-1 system, which consists of a very small dwarf star and seven small planets, several of which are in the habitable zone (figure 8.11). Among the exoplanets' candidates for habitability, some objects are exo-Earths (Kepler-438 b, Proxima Centauri b, TRAPPIST-1 e, TRAPPIST-1 g, Gliese 667 C c, Kepler 186 f), the others are super-Earths. Proxima Centauri b is particularly interesting because it orbits around the closest star to Earth, Proxima Centauri. It is nevertheless located more than 4 light-years away from us.

FIG. 8.11 – The TRAPPIST-1 system, consisting of a very small dwarf star and 7 planets of mass comparable to that of the Earth (only their relative sizes are meaningful). The whole system is much smaller than the orbit of Mercury, as shown below. Adapted from Lequeux *et al.*, *La révolution des exoplanètes*, L'Astronomie, 2018.

But beware! Detecting a rocky exoplanet in the habitable zone of its star is only the first step... Other factors, more difficult to quantify, must necessarily intervene in the process leading to the appearance of life: the ellipticity of the orbit which can create strong variations in the stellar radiation received; the obliquity of the planetary axis, responsible for the seasons; the presence or not of a magnetic field which tends to protect the planetary atmosphere from the sometimes violent effects of the stellar wind; the type of star, which can show violent bursts of activity, etc.

In an attempt to clarify these notions, several research teams have defined indicators that are supposed to evaluate the habitability of an exoplanet. Several approaches are possible, depending on the criteria retained: these can be the similarity with the Earth, the preferred energy source (thus the spectral type of the host star), the similarity with the Earth's atmosphere, the presence of liquid water and carbon. Despite their necessarily arbitrary character, these indicators can be useful to make a first selection among the thousands of exoplanets know today.

The first indicator that astronomers have considered is a quantity that has been used by biologists for several decades: the *Habitability Suitability Index* (HSI), which varies between 0 and 1. However, it has proved difficult to apply to exobiology. Another index better adapted to what we know about exoplanets is their degree of resemblance to the Earth: the *Earth Similarity Index* (ESI). Also comprised between 0 (no similarity) and 1 (the case of the Earth), this factor takes into account the physical characteristics (radius, density, escape velocity, surface temperature). About twenty exoplanets have thus been assigned an index greater than 0.5, with Venus and Mars receiving values of 0.56 and 0.78, respectively. However, it is clear that this index is not a good criterion for habitability: indeed the Moon has an index of 0.57 because of its physical analogies with the Earth, while we know that all life is impossible on it.

The best indicator is finally the *Planetary Habitability Index* (PHI) which varies from 0 (lack of habitability) to 4.67 (the case of the Earth). Instead, we use its value normalized with respect to the Earth, which varies between 0 and 1. Defined by Dirk Schulze-Makuch and his team in 2011, it takes into account mass, density, the presence of a surface, a neutral atmosphere, a magnetosphere, the presence of oxidants, reducing agents and organic compounds, the existence of a solvent (liquid water or other), the nature of the energy source. The exercise can be further complicated by defining the *Biological Complexity Index* (BCI). This involves, in addition to the above factors, geophysical and orbital considerations, as well as the age of the system.

In conclusion, the concept of habitability indicator is interesting in itself, in the sense that it can allow a first choice in the vast catalog of exoplanets. However, it does not allow us to go any further for the moment, since most of the quantities involved are currently unknown.

8.7   Satellites Around Giant Exoplanets?

We have seen that some external satellites of the Solar System offer interesting potential niches in terms of habitability (see chapter 7). Could we encounter the same type of object in exoplanetary systems? *A priori*, the answer is yes: the same mechanisms responsible for the formation of the regular satellites of the giant planets in the Solar System could be at the origin of icy "exosatellites" comparable to those we know. These objects could also constitute privileged niches for exobiology. Unfortunately, it would be very difficult to discover any underwater life under their icy surface! Another case would appear more interesting, that of possible satellites in orbit around "meso-exoplanets" located in the habitable zone of their star. These could then have a mass comparable to that of the Earth and shelter liquid water. The search for traces of life on these satellites would then join the search for life on rocky exoplanets.

Have we discovered today satellites of exoplanets? In the recent years, a few candidates have been proposed. In 2013, a team of researchers announced the discovery by the micro-lensing method of a couple composed of a giant exoplanet and a sub-terrestrial mass satellite; however, the system could also be composed of an exoplanet around a brown dwarf. In 2018, a new discovery was announced, based on transit observations carried out with the Hubble Space Telescope: a satellite of mass comparable to that of Neptune, orbiting an exoplanet of several Jovian masses, Kepler-1625 b. With the constant refinement of techniques for the detection and analysis of exoplanets, we can expect the discoveries of "exo-moons" to increase rapidly in the years to come.

8.8 How to Determine the Atmospheric Composition of an Exoplanet?

What measurements would allow us to better understand the nature of an exoplanet? Of course, if we were able to make a visible image of it, as has been done for objects in the Solar System, we would be able to detect the presence of continents, clouds, ice caps, etc. These observations will probably come in their time, but given the current means, they are not for tomorrow: remember that in most cases, the exoplanet is identified indirectly from the behavior of the star (motion or periodic decrease in luminosity) and is not directly visible. However, astronomers have a very powerful tool to study the atmospheric composition of an exoplanet observed by transit: the study of its spectrum during this transit. To measure this spectrum, one proceeds by difference between the spectrum observed during the transit and that observed just before or just after.

Until now, we have considered only one kind of transit: the one that occurs when the exoplanet passes in front of its star; it is the one that produces the strongest decrease of the stellar light, and therefore the most able to detect an exoplanet; it is called "primary transit". In this configuration, the atmosphere of the exoplanet is observed in transmission in front of the stellar light, as a ring surrounding the disk of the exoplanet, like Venus passing in front of the Sun. The atmosphere is then observed at the terminator, *i.e.* at the transition between the day side and the night side of the planet, and we see the absorption spectrum of this atmosphere. But there is another possible observation, when the exoplanet passes behind its star (the "secondary transit"), the difference between the signals observed before and after the

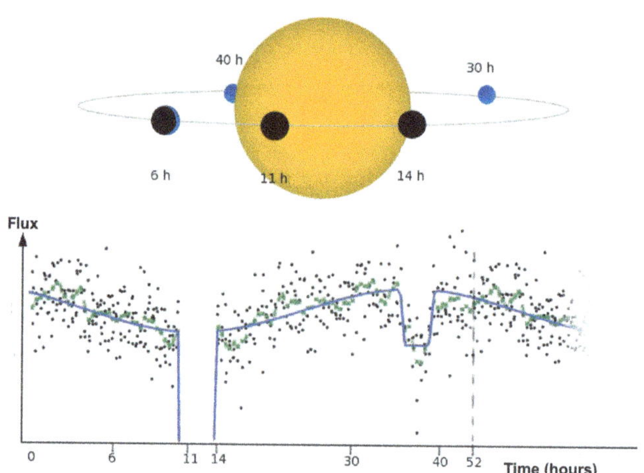

FIG. 8.12 – Diagram of the light curve of a star when its planet passes in front of it (primary transit) or behind it (secondary transit). The revolution period of the planet is 52 h. Outside the transits, the light curve evolves as a function of the illuminated fraction of the planet that is added to the star's light. Adapted from Lequeux *et al.*, *La révolution des exoplanètes*, 2017, EDP-Sciences.

second transit and during it is smaller; we then observe the proper radiation emitted by the day side of the exoplanet (figure 8.12).

From the beginning of the 2000s, following the success of the transit method, the researchers, not content with measuring the light coming from the exoplanet, in transmission during the primary transit or in emission during the secondary transit, have undertaken to make its spectrum. These measurements are extremely difficult: it is a question of measuring the spectral variations of a quantity representing, at best, some 0.01% of the stellar light! This precision requires, in most cases, observations from space. The first measurements, on giant exoplanets (gaseous exoplanets) in orbit around bright nearby stars, were made during the primary transit in the visible or ultraviolet range thanks to the *Hubble Space Telescope* (*HST*). They allowed the detection of various species in atomic form (hydrogen, helium, carbon, sodium, potassium...). The observations were then extended to the infrared range using the *HST* but also *Spitzer*, an infrared space telescope launched in 2003 by NASA. The infrared range allows the observation of most molecules from the study of their vibration or rotation transitions (see chapter 2). Most of the small atmospheric molecules observed in the Solar System have been detected in exoplanets, starting with water H_2O, but also methane CH_4 (figure 8.13), carbon dioxide CO_2 (figure 8.14), carbon monoxide CO and ammonia NH_3.

Note that the most abundant terrestrial atmospheric gases, nitrogen N_2 and molecular oxygen O_2, are devoid of any significant spectral signature, due to the absence of a dipole moment in these symmetrical molecules, and are therefore difficult to observe. The fact that these molecules have not been detected does not imply that they are absent in the exoplanets. Transit spectroscopy of exoplanets has

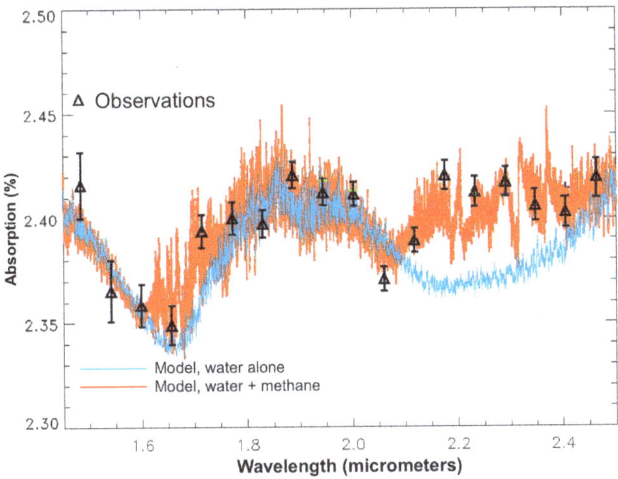

FIG. 8.13 − Near-infrared spectrum obtained during the primary transit of the giant exoplanet HD189733 b in front of its star. The quantity in ordinate is the percentage of absorption of the stellar flux; the more abundant a molecule is, the higher this quantity is. The triangles represent the observations obtained from the space telescopes *HST* and *Spitzer*. The curves represent models, including water vapor H_2O alone (in blue) or water vapor plus methane CH_4 (in orange). The good agreement of the orange curve with the measurements indicates the presence of both molecules in the atmosphere of the exoplanet. Adapted from Swain *et al.*, *Nature* 452, 329 (2008).

progressed considerably over the last fifteen years: at present, we have information on the atmospheric composition (atoms or molecules) of about fifty exoplanets, all giants, generally very hot. Unfortunately, no spectrum of a terrestrial-mass planet could be obtained up to now. This field of research is developing thanks to the *CHEOPS* mission of ESA, launched in December 2019. The *James Webb Space Telescope* (*JWST*), the successor of the *HST*, due to launch in 2021, might obtain spectra of low-mass exoplanets.

8.9 How to Search for Life from the Spectrum of an Exoplanet?

Let us return to our first objective, the identification of traces of life on an exoplanet. Can we detect life from the measurement of its spectrum? Let us imagine that we observe the Earth from a nearby star. Could we derive the presence of life from its atmospheric composition? We know that on Earth, the production of molecular oxygen, through chlorophyll photosynthesis, is a consequence of the development of life (see chapter 6). In the absence of other elements, we can follow this track, although this approach is clearly anthropocentric. Atmospheric chemists agree on

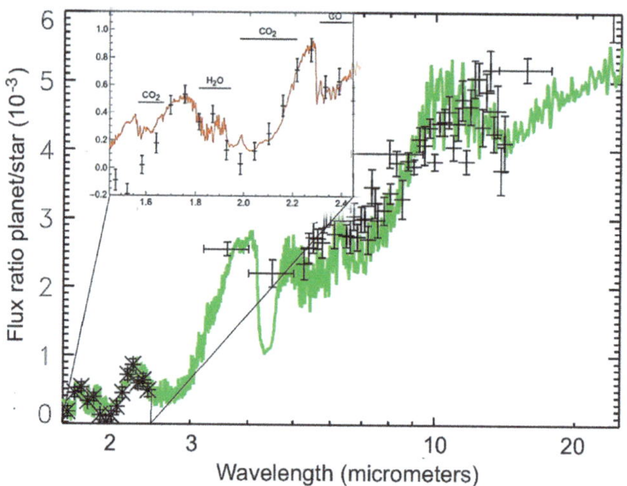

FIG. 8.14 – Near and mid-infrared spectrum of the giant exoplanet HD189733 b (the same as for figure 8.13) obtained during the passage of the object behind its star (secondary transit). The quantity in ordinate is the ratio of the flux of the exoplanet to that of the star. It increases with wavelength because the planet is colder than the star. Measurements in the near infrared (top left box) come from the *HST*, the others from the *Spitzer* space telescope. The comparison of the data (in black, with error bars) with the models (in color) makes it possible to constrain the atmospheric composition of the exoplanet. Adapted from Tinetti and Griffith, *ASP Conference Series* 430, 115 (2010).

the following fact: the presence of large quantities of oxygen in an atmosphere requires processes involving life; it is true that in the atmosphere of Mars and Venus (see chapter 2), oxygen is present only in trace amounts.

The only thing left to do is to try to discover the oxygen signature in the spectrum of an exoplanet. Unfortunately, as we have seen, this molecule is not very spectroscopically active. It does have a transition in the visible range, around 0.7 μm, but this transition is weak and is in a congested spectral range. There is a better lead: the search for ozone O_3. Ozone is derived from O_2 by photochemistry and forms on Earth a stratospheric layer that protects us from solar ultraviolet radiation. Compared to oxygen, the O_3 molecule has the large advantage of having, in the infrared, a very strong spectral signature around 9.6 μm (figure 8.15). This transition will be sought in priority on exoplanets potentially habitable by the *JWST* when it enters into service (the use of an artificial satellite is mandatory because of the existence of terrestrial ozone). Taking their research further, biochemists have defined a list of "biomarkers": these are molecules whose presence is expected in the presence of life that would have developed as on Earth. They include, of course, molecular oxygen and ozone, but also the simultaneous presence of methane, carbon dioxide and water.

Another possibility, also directly derived from our experience of life on Earth, is the detection of chlorophyll whose spectrum shows a transition at around 0.7 μm. It

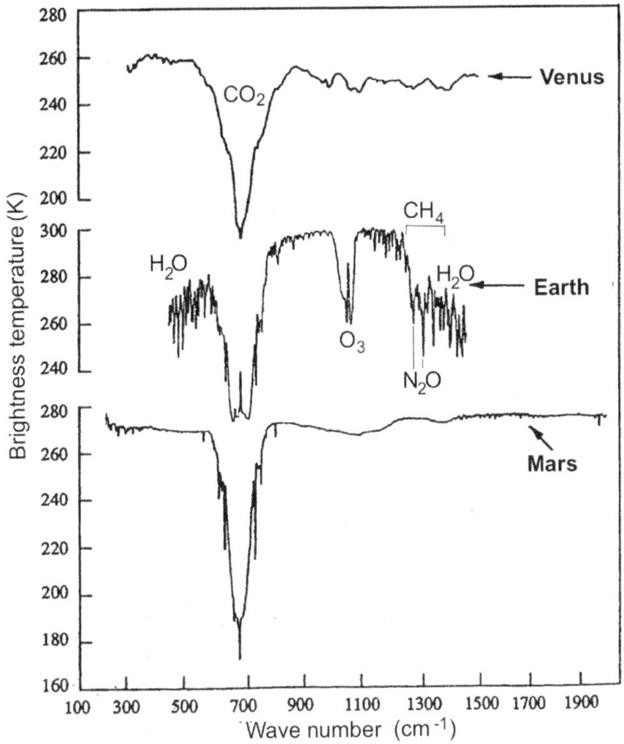

FIG. 8.15 – Infrared spectra of the three terrestrial planets Venus, Earth and Mars between 5 and 33 μm (300–2000 cm^{-1}). The quantity on the ordinate is the cloud temperature (in the case of Venus) and the surface temperature (in the case of Earth and Mars). The three spectra show the spectral signature of carbon dioxide CO$_2$ at 660 cm^{-1} (15 μm). The spectrum of the Earth shows in addition the signatures of water H$_2$O, nitrous oxide N$_2$O and methane CH$_4$ beyond 1300 cm^{-1}. The signature of ozone O$_3$, absent from the spectra of Venus and Mars, also appears very clearly in the Earth spectrum at 1040 cm^{-1} (9.6 μm). It is considered a distinctive sign of the presence of life on Earth. From Hanel *et al.*, *Exploration of the solar system by infrared remote sensing*, Cambridge University Press, 1992.

corresponds to a slope break of a few percent in the spectrum of the Earth, which has been observed (figure 8.16). Its detection, already very difficult from the Earth because of the cloud cover that masks the continents in places, would be even more difficult in the case of an exoplanet.

In the longer term, it might be possible to obtain direct images of terrestrial exoplanets or super-Earths in the visible and near-infrared. Two types of methods might be used. The first is direct imaging using adaptive optics techniques to correct defects related to the Earth's atmospheric transmission and coronagraphy to eliminate light from the host star. This method is already at work on 10-m class telescopes and allows direct imaging of giant exoplanets when they are far enough away from their star, and spectroscopy is already possible. By transposing it in

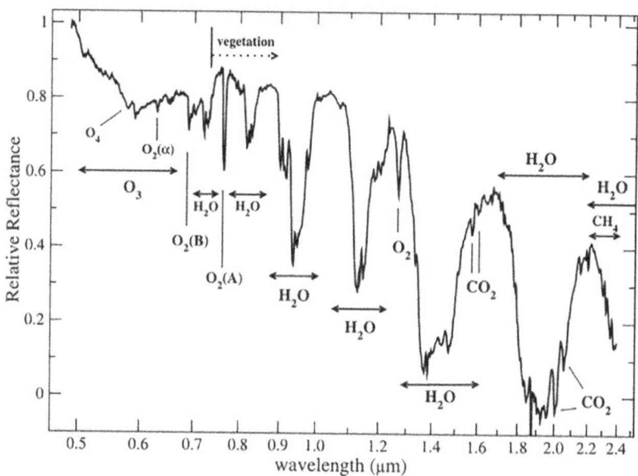

FIG. 8.16 – The spectrum of ashy light in the visible and near infrared. This is the sunlight reflected by the Earth on the dark side of the Moon, observed from the Earth. The discontinuity due to the presence of chlorophyll is visible at 0.72 µm ("vegetation"), but would be difficult to identify on an exoplanet because of the presence of atmospheric molecules, especially water vapor. We note the signature of oxygen bands at 0.76 and 1.27 µm. Adapted from Turnbull *et al.*, *Astrophysical Journal* 644, 551 (2006).

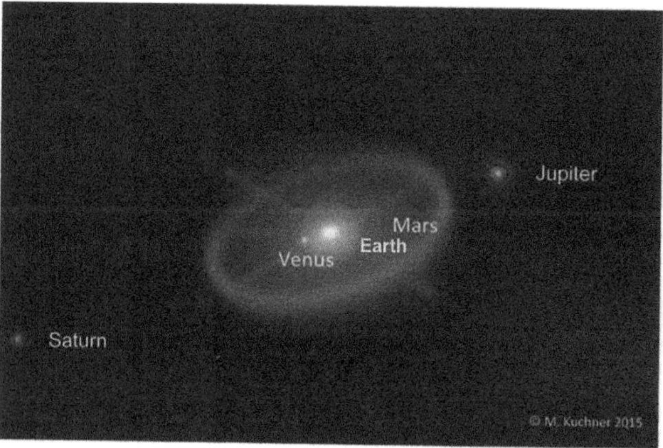

FIG. 8.17 – The Inner Solar System as it would appear from a distance of 40 light-years, with a telescope of 12 m diameter equipped with a large coronagraph of 100 m diameter, interposed at great distance © M. Kuchner and NASA *LUVOIR* project.

space, it would be theoretically possible to image a system equivalent to the Solar System which would be located 40 light-years away: it is the *LUVOIR* (Large Ultraviolet Optical Infrared Surveyor) project currently under study at NASA (figure 8.17). The second method would consist in sending a constellation of telescopes operating in interferometric mode. According to the hyper-telescope concept developed by the astronomer Antoine Labeyrie, a flotilla of about a hundred telescopes a few meters in diameter could thus produce images of exoplanets. It would then be possible to study their morphology. Such projects will take time, but we are not in science fiction: the technologies exist or are in the process of being developed, and such space missions could come into being in the decades to come. We will then be able to progress further in the exploration of these new worlds that offer exobiology an infinite number of possibilities.

Chapter 9

Conclusions: Some Future Directions in Exobiology

Our overview of possible sites of extraterrestrial life made us discover several possible tracks. The first one, the closest to us, concerns the planet Mars, on which we still hope to discover one day fossil or even present forms of life. The second, which we have only briefly mentioned, concerns certain satellites of the giant planets of the Solar System that could shelter, under the icy crust that forms their surface, an ocean of liquid water, perhaps in contact with the rocky core present in their interior. In the present state of our knowledge, one can only hope to find, in both cases, very primitive forms of life; within the outer satellites, one could at best expect to find the forms of life present on Earth in the oceans. On the other hand, the exploration of rocky exoplanets opens up an infinite range of possibilities in terms of exobiology. If life may have appeared and developed on some of them, nothing prevents us from imagining that, in some cases, it may have reached a degree of civilization equal or superior to ours... A new question then arises: beyond the detection of these possible forms of life, could we one day envisage a possible way of communicating with these distant civilizations? What has been for more than a century a theme dear to science fiction has become, in recent decades, a real scientific question.

9.1 The Future of Mars Exploration

Let us go back to the Solar system. After eight years of *in-situ* exploration by the American rover *Curiosity*, what have we learned? Since it began operations at the Bradbury site inside Gale Crater, the rover has traveled more than 21 km, while climbing the flanks of Mount Sharp at an altitude of about 300 m. Perhaps the most striking result of this campaign is the discovery of an ancient lake and the chemical ingredients necessary for the development of life, which makes it possible to describe Gale Crater as a "habitable site". The *Curiosity* mass spectrometer also revealed a small quantity of methane (less than 1 ppbv) which seems to show seasonal

DOI: 10.1051/978-2-7598-2563-9.c009

variations, to which are superimposed possible transient "plumes" up to 7 ppbv, which could result from an episodic outgassing of the sub-surface, whose origin remains to be understood. However, we have seen that the *Trace Gas Orbiter* has found no trace of methane at an altitude above 5 km, so this result is still a matter of controversy. The search for organic matter has long been negative. Organic molecules, associated with the presence of sulfur, were finally discovered in two very old samples, dating back 3.5 billion years; the presence of sulfur could have contributed to their preservation. The balance is therefore mixed: an "habitable" environment more than 3 billion years ago, but no trace of fossil life.

What will be the future of Martian exploration? For decades, scientists have been advocating the return of Martian samples. As was brilliantly shown in the 1970s, the study of lunar samples brought back by space missions, and their analysis in the laboratory, using the most sophisticated electronic scanning microscopes and ion probes, has made it possible to precisely determine their elemental composition and to date them. In the case of Mars, this analysis could be carried out partially, thanks to the Martian meteorites. According to the NASA records of 2018, over a hundred of these objects have been discovered on Earth; most of them are named "SNC", in reference to the place of origin of the first three of them (Shergotty in India, Nakhla in Egypt and Chassigny in France). The analysis of the isotopic composition of oxygen, different from that of the Earth and compared to that measured in the atmosphere of Mars by the *Viking* probes, allowed to determine their origin without ambiguity. Where do they come from? They must have been ejected from the depths of the Martian crust following the impact created by a giant meteorite. In 1979, the ALH84001 meteorite made the headlines: discovered in 1984 in Antarctica, it presented structures, under the microscope, that had similarities with fossilized bacteria. After a long controversy, it is generally accepted that these structures come from a subsequent contamination of the meteorite by the terrestrial environment.

Therefore, since we now have several tens of kilograms of Martian samples, why would we want to bring others back to Earth? Because the samples collected on Earth have been ejected from Mars by meteoritic impacts; sedimentary rocks, where we would could hope to find fossils or traces of biogenic materials, do not resist the thermal shock generated by these impacts. A collection of carefully selected samples would provide access to surface materials of different types, which would bear traces of various processes such as sedimentation or weathering of rocks by water or the atmosphere, or even possible biological signatures... Ideally, samples should be collected from several sites: lava flows to define a Martian chronology, polar ice cores to trace the paleoclimate of Mars, sedimentary materials to search for possible traces of life.

The first discussions on a mission to return Martian samples date back to the 1970s, following the progress made on the dating of the Moon thanks to lunar samples. But the project suffered from the disaffection of the general public following the negative results of the *Viking* probes on the presence of life on Mars. Other projects emerged in the following decades, but all of them ran into the obstacle of the cost of the mission (several billion dollars). Following the technological success of the *Curiosity* mission, NASA decided in 2015 to develop a new rover derived from *Curiosity, Mars2020,* whose mission will be to collect a series of samples.

This mission was launched in July 2020 and has delivered a robot, called *Perseverance*, at the surface of Mars, on 18 February 2021. Other modules, operated and funded by ESA and other agencies, are expected to collect these samples and return them to Earth by 2030. Operated by Russia and ESA, the *ExoMars 2022* program should deliver another rover, equipped with a suite of instruments similar to those of *Curiosity* and *Perseverance*. In addition, this rover should be able to drill in the soil down to a depth of two meters.

The study team in charge of defining the *Mars Sample Return* program has proposed, in 2018, a scenario including at least three launches to bring back 500 g of Martian samples (figure 9.1). The first launch is for a rover designed for selecting the most interesting samples, introducing them into a special container; this role will be played by the *Mars2020 Perseverance* mission. Another launch will bring a rover to collect these samples, which will be sent in orbit around Mars with a small rocket; in a last step, this container will be captured by an orbiter around Mars and brought back to the Earth. This scenario, not funded yet, is very complex, and the various steps require more technical developments before they become real. In particular, the development of a rocket in the Martian environment is not easy, in view of its oxidizing properties and its strong thermal variations.

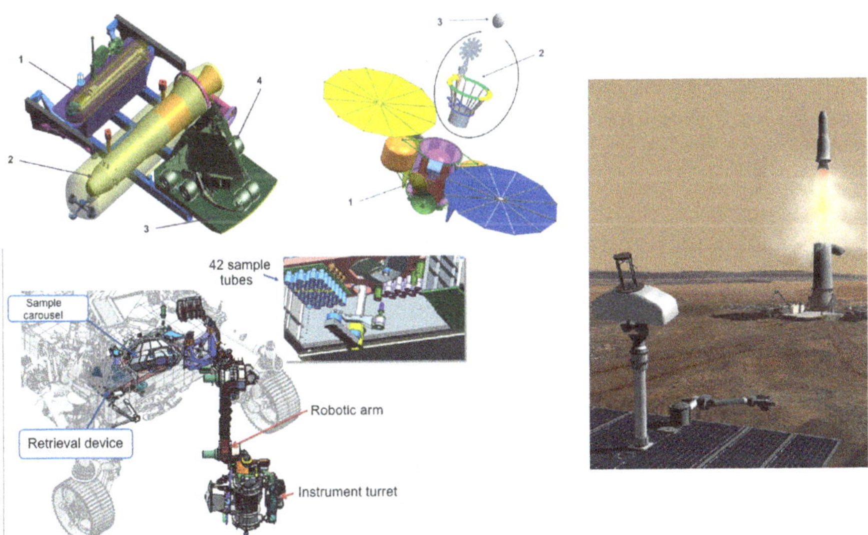

FIG. 9.1 – The Martian sample return scenario currently considered by NASA and ESA. The *Mars2020* rover (*Perseverance*, bottom left), developed by NASA, is in charge of collecting samples and arrived on Mars in February 2021. All the other elements are currently under study at ESA, but are neither finalized nor financed for the moment. A first vehicle (top left) would carry an articulated arm to collect the samples (1), the rocket intended for their takeoff (2), a launch platform (3) and the rover in charge of collecting the samples (4). A Martian orbiter (top, middle), would include a capsule to be returned to the Earth (1) with the recovery system (2) of the sample container (3). Adapted from Wikipedia Commons © NASA.

Suppose we finally have Martian samples collected *in situ* and brought back to Earth. Before these are available to the scientific community, they will first have to stay for several weeks in a "curation center" set up for this purpose under the responsibility of the major space agencies. This constraint is imposed by the Planetary Protection Protocol set up by COSPAR (Committee on Space Research), an international organization coordinating space research. The objective of planetary protection is, on the one hand, to avoid contaminating with terrestrial micro-organisms the extraterrestrial environments explored by the space probes (this is why they the probes and rovers are sterilized as well as possible before their launch), and on the other hand to avoid a possible contamination of our planet by extraterrestrial samples during their return to Earth; this is the purpose of their quarantine in a dedicated curation center. In the case of lunar samples, this curation center was in the United States; in the perspective of an international mission to return Martian samples, initiatives are being put in place to develop one of these centers in Europe.

Can we imagine techniques to characterize living matter from the analysis of Martian samples? A possible diagnosis is based on the isotopic ratio of carbon, $^{12}C/^{13}C$. On Earth, this ratio, measured in minerals, is equal to 90. However, in living organisms, it is altered by the effect of photosynthesis. Indeed, plants fix carbon by means of two photosynthetic cycles leading to the formation of molecules with 3 or 4 carbon atoms. In both cases, they use ^{12}C carbon preferentially when they convert carbon dioxide into organic matter, which leads to an enrichment of the $^{12}C/^{13}C$ ratio of 2.4% in living matter. Measuring such enrichment would *a priori* be possible in Martian samples; however, this criterion limits the research to life forms associated with photosynthesis, and does not apply to other life forms.

Another possible diagnosis to characterize living matter is the study of chirality. What is it about? Many organic molecules have two isomeric forms, symmetrical with respect to a mirror; this is the case, for example, of a molecule composed of a carbon atom surrounded by four different atoms or four different groups of atoms. These two isomers, crossed by a light beam, have the property of rotating the plane of polarization of this light in opposite directions; they are called D (for dextrorotatory) or L (for levorotatory). However, it so happens that the chiral molecules of terrestrial organic matter all chose a given type: this is what we call the homochirality of living matter. For example, the amino acids of proteins are all of type L, the sugars of type D, etc. What could be the cause of this? Recent laboratory studies have shown that the synthesis of amino acids irradiated by a polarized ultraviolet flux (as produced for example in a region of massive star formation) leads to an excess of L-chirality, comparable to what is measured in meteorites. The prebiotic molecules at the origin of life on Earth could have exhibited an excess of L following this mechanism; the L chirality would then have been imposed on Earth by natural selection on the whole living world. The study of the chirality of organic molecules is done by classical optical polarimetry techniques; it will be part of the laboratory analysis of future Martian samples.

9.2 How to Detect Traces of Life *In Situ?*

The technique of measuring the $^{12}C/^{13}C$ ratio in gas or rock samples taken *in situ* can also be used in Martian robotic missions, if they are able to detect organic matter. For this reason, the *Curiosity* rover has been equipped with a tunable laser spectrometer (the instrument that led to the tentative detection of methane), capable of measuring the abundances of $^{12}CH_4$ and $^{13}CH_4$ separately. Unfortunately, the quantities of methane detected by *Curiosity* so far are not sufficient to allow this measurement. Its principle is taken up again for the *in situ* analyses that will be carried out by *Curiosity*'s successor, *Mars2020 Perseverance*, and the rover *ExoMars 2022*.

Apart from *Curiosity*'s and *Perseverance*'s tunable laser spectrometers, no living matter research experiment has been embarked on a Martian mission since the *Viking* lander adventure. At that time, the experiments were oriented towards the search for metabolic activity, *i.e.* the demonstration of chemical reactions by which a living organism exchanges with the surrounding environment. Due to the difficulty of interpreting the *Viking* results, this type of experiment was not repeated. The perspectives are rather oriented towards the search for "biomarkers", *i.e.* signatures of a past or present life. The definition of biomarker is here broader than the one we used in the case of the analysis of the spectrum of exoplanets (chapter 8): it is not only a matter of searching for ozone, carbon dioxide or methane in the Martian atmosphere, or chlorophyll on the surface of an exoplanet, but of identifying specific molecules (amino acids, proteins, nucleic acids, and even RNA or DNA), by means of "biochips" specific to each molecule or each type of molecule. A British experiment of this type, *LMC* (*Life Marker Chip*) was thus selected as part of the payload of the descent module of the *ExoMars2022* mission, but it had to be removed due to lack of available space. On the other hand, a more classic American experiment for the search for organic molecules, *MOMA* (*Mars Organic Molecule Analyzer*), based on chromatography and mass spectroscopy (like the *Curiosity* rover experiment), will be part of the *ExoMars2022* payload.

9.3 Towards an Inhabited Exploration of Mars?

In all of the above, we have considered planetary (and a fortiori exoplanetary) space exploration from the perspective of robotic missions. Indeed, these missions are responsible for the prodigious progress of knowledge achieved for over fifty years in astronomy, whether about the knowledge of our Solar System (thanks to dedicated planetary missions) or that of the Universe (thanks to space observatories operating in the entire wavelength range of electromagnetic waves).

However, the sending of men to the Moon, between 1969 and 1972, launched another line of thought: that of inhabited exploration of the nearest planets. Given the hostile conditions presented by Mercury and Venus (absence of atmosphere and high temperature on the day side of Mercury; colossal pressure, clouds of sulfuric acid and torrid atmosphere on the surface of Venus), the choice was quickly made for the planet Mars. Prior to *Apollo 11*, inhabited exploration of planets was science fiction. Since the landing of men on the Moon, it has given a new impetus to the dreams of space conquest conveyed in the literature for millennia, starting with the myth of Icarus, and particularly flourishing in recent centuries; one need only cite the voyages to the Moon evoked by Cyrano de Bergerac, Jules Verne or H. G. Wells. In fact, the question is more philosophical than scientific: by escaping the Earth's gravity, humans escape their condition as earthlings and push back the boundaries of their environment. The old concept of colonization, which accompanied all the stages of the great discoveries, is back in force, especially since we now know that "little green men" do not exist on Mars...

In addition to the public's enthusiasm for inhabited exploration, the economic and political context has played an essential role. Without the confrontation of two superpowers, the United States and the Soviet Union following the Second World War, manned exploration of the Moon would never have taken place; once this challenge was over, the *Apollo* program was quickly interrupted, leaving the way open for planetary robotic exploration; science made the most of it. Today, the situation is changing. The emerging powers, for both political and economic reasons, also want to participate in the conquest of space, with, initially, the sending of a man into space, then the return of manned exploration of the Moon. This is the case in particular of China, which in 2018 deposited the robot *Chang'e 4* on the far side of the Moon; the *Chang'e 5* probe, launched in November 2020, is expected to bring lunar samples back to Earth, and China's more distant goal is the installation of a lunar base. The economic environment is also changing, with the appearance of new private actors. The Space X company has developed a new concept of access to space by low-cost reusable spacecraft, the Falcon rockets. Under contract from NASA, Space X has been carrying out supply flights to the *International Space Station* (ISS) since 2012 and has been transporting astronauts to the ISS in 2020. The director of Space X, Elon Musk, has the stated goal of transporting settlers to the planet Mars. The cargo and soon-to-be-inhabited service of the ISS is also carried out by Boeing, an older industrialist in the sector. The company Blue Origin, whose CEO is Jeff Bezos, founder of Amazon, is also developing reusable rockets to take tourists into space, and has also proposed a concept for lunar landing. This new context forces all the major space agencies to study projects of inhabited space exploration, whether it is for the Moon, Mars or even an asteroid.

What about science? Let us be clear: science is not the driving force behind this approach, although it could, of course, benefit from it. We have seen that the study of lunar samples has been decisive in establishing the dating of Solar system objects; however, their return did not require the sending of a manned mission. If the *Apollo* missions represented a fantastic human adventure, their cost was no less exorbitant; this was the cause of the interruption of this program, once the political stakes have been overcome. The scientific success of robotic missions can be explained by the ability to miniaturize the measuring instruments while developing their performance ever further. We see it today in the field of medicine in particular: robotics, combined with artificial intelligence, is increasingly capable of reproducing human gestures and behavior. In contrast, we will never be able to miniaturize the human body!

So what should we think of the manned flight program to Mars? For a cost one hundred to one thousand times that of a robotic mission, the scientific return would not be truly transformed. Remember that the cost of a robotic mission to the surface of Mars (*Curiosity* for example) is a few billion dollars. The cost of the *Apollo* program, infinitely less complex than a manned mission to Mars, was about 170 billion dollars. Perhaps manned exploration of Mars, carried out for political and economic reasons, will take place in several decades; it will nevertheless be a financial drain, at a time when all energies should be directed towards the survival of our own planet.

9.4 Towards External Satellites, Other Possible Niches for Life

We have briefly mentioned the prospects for exobiology inside some satellites of giant planets. Thanks to the tidal effects generated by the presence of their host planet, the temperature inside these objects could allow the presence of water in liquid form, thus opening new perspectives in terms of exobiology. Two of them appear particularly promising: the Galilean satellite Europa, in orbit around Jupiter, and Saturn's small satellite Enceladus. Indeed, the ocean of salt liquid water that is probably present under their icy surface could, according to internal structure models, be in direct contact with the silicate nucleus of these objects.

Ideally, we would need to send a probe inside these oceans whose presence, however, is not yet demonstrated. We are still a long way from this: let us remember how many years it took to carry out the EPICA sounding in Antarctica down to a depth of three kilometers... and in the case of the external satellites, we do not know the thickness of the ice crust! But the first stages of exploration are taking place. The first target will be Europa, twice as close to us as Enceladus. The main difficulty is the intense radiation of the particles trapped in Jupiter's magnetic field, which has the effect of very quickly damaging the instruments of nearby space probes. Two space missions are planned and under development. The European mission *JUICE* (Jupiter and Icy Satellite Explorer), whose launch is planned for 2022, should arrive in 2030 for a series of flybys of the three Galilean satellites furthest from Jupiter (including two close flybys of Europa), followed by an orbit insertion around Ganymede in 2032 (figure 9.2). The American *Europa Clipper* mission, under development at NASA, plans a launch between 2022 and 2025, an arrival in Jupiter's system a little more than six years later, and a series of 45 close flybys of Europa. The first objective will be to confirm the existence of the liquid water ocean, determine its characteristics and study the exchange processes between the ocean and the surface. Finally, let us not forget Saturn's system. Titan, its largest satellite, a prime target of the *Cassini–Huygens* mission conducted jointly by ESA and NASA, has an atmosphere rich in prebiotic molecules, which presents analogies with the primitive Earth. After the spectacular success of the *Cassini* mission and its successful completion in September 2017, many planetary scientists dream of continuing Titan's exploration.

FIG. 9.2 – The *JUICE* (*JUpiter ICy moons Explorer*) mission, selected by ESA and currently under development for a launch in 2022, aims at exploring the moons of Jupiter and their habitable conditions. The spacecraft will make several flybys over Europa, Ganymede and Callisto and will then orbit Ganymede. This artist's view shows the probe in orbit around Ganymede, whose internal structure is schematized with the underground ocean. Europa is on the bottom left in the foreground and Jupiter is in the background with its satellite Io. The Callisto satellite is shown below the probe. © ESA.

For both Titan and Enceladus, there are exploration plans from the satellite's orbit. *Dragonfly* is a NASA mission devoted to the exploration of Titan with a rotary-wing aircraft planned for multiple takeoffs and landings in order to analyze different sites. Its main objective is astrobiology and extraterrestrial habitability. The mission, selected in 2019, is planned for a launch around 2027 and an arrival on Titan around 2036. The scientific payload includes a suite of cameras, a mass spectrometer, a gamma-ray spectrometer and a seismometer, for a planned science phase over two years.

9.5 Exploring Exoplanets: The Prospects

As soon as we leave the Solar System, the debate takes another turn. It is no longer a question of getting closer to the objects we are interested in, but of studying them from a distance, or even trying to communicate with them by means of electromagnetic signals. Let us recall that the nearest star, Proxima Centauri, is located more than 4 light-years from the Sun. No technology today makes it possible to imagine sending any sophisticated interplanetary probe to such a distance in a time compatible with the human lifetime. For a few decades (and probably more), the exploration of exoplanets will require their remote study, using ground-based or near-Earth telescopes equipped with increasingly sophisticated measuring instruments.

In the short and medium term, the roadmap is well defined. First, we need to build a true classification of exoplanets whose astonishing diversity, both physical and orbital, we have discovered; to carry out this census, we must continue exoplanetary exploration by all possible techniques since they give access, as we have seen, to different classes of objects. In parallel, we must improve our knowledge of the nature of the different classes of objects by determining their atmospheric composition. The variety of exoplanets is such that we can expect to discover atmospheres with extraordinary characteristics in terms of physical conditions and chemical composition. It will then be necessary to understand, among the "anomalies" observed, those that are likely to betray the presence of a life form.

In the field of exoplanet detection, the astrometric satellite *Gaia* (figure 9.3), launched by ESA in 2013, is going to change the landscape over the next decade. This mission is measuring, with unprecedented accuracy, the positions and velocities of more than a billion stars. In addition to the fundamental contribution that these data will provide to stellar and galactic physics, *Gaia* should finally enable the detection of exoplanets by direct astrometry, which Peter Van de Kamp attempted unsuccessfully a century ago. The discovery of more than 20 000 exoplanets is thus expected, with, for the brightest of them, the determination of their mass and orbital parameters. In addition, *Gaia* will also perform planetary transit measurements that should lead to the detection of more than 6000 exoplanets. Within a few years, the catalog of exoplanets discovered should be multiplied by a factor of five or more.

To these results will be added those of the space missions dedicated to planetary transits, in the wake of *CoRoT* and *Kepler*: the American *TESS* (*Transit Exoplanet Survey Satellite*) mission, launched by NASA in April 2018. Dedicated to the observation of exoplanets around bright stars, it should enable the detection of more

Fig. 9.3 – Launched by the European Space Agency in 2013, the *Gaia* mission is an astrometry mission designed to measure the positions and velocities of more than a billion stars in our Galaxy with unparalleled accuracy. The second *Gaia* catalog was released in April 2018, a preliminary version of the third one has been published in December 2020, and its full version will be available during the first semester of 2022. These data should allow the discovery of many new exoplanets. © ESA.

than a thousand exoplanets which could then, unlike *Kepler*, be the subject of velocimetry and spectroscopy studies to better determine their physical characteristics. On the European side, the *CHEOPS* mission, developed in partnership between Switzerland and ESA, will also be dedicated to the "transit analysis" of medium-sized exoplanets (super-Earth or mini-Neptune); it was launched in December 2019. Both are working satisfactorily. At the same time, networks of small automatic telescopes from Earth are multiplying. This is the case of the *NGTS* (*Next Generation exoplanet Transit Survey*), a set of twelve 20-cm diameter telescopes located in Chile, at the ESO site on Mount Paranal.

Astronomers did not abandon the velocimetry method, which proved so successful in discovering the first exoplanets. Following in the footsteps of the pioneering spectrometers *CORALIE* and *HARPS*, the *ESPRESSO* instrument has been in service since November 2017 at the *VLT* (*Very Large Telescope*), on ESO's Paranal site in Chile. It has the capability of working with a single telescope or combining light from four 8-m telescopes; it can also operate in interferometric mode with the four auxiliary telescopes of the VLT to obtain the best possible angular resolution, and thus very precise astrometry. The same type of instrument is also under study to equip the ESO *ELT* (figure 9.4) at the end of the decade 2020.

FIG. 9.4 – The *Extremely Large European Telescope* (*ELT*), developed by ESO, whose model is shown here, is a 39-m diameter telescope currently under construction at Mount Armazones, Chile. It is the largest telescope in the world under construction to date. The first light is expected at horizon 2025. © ESO.

Another field with a bright future is direct imaging of exoplanets, thanks to adaptive optics and coronagraphy techniques that allow us to eliminate the light of the central star. The entry into service of Extremely Large Telescopes with diameters ranging from 30 to 40 m, equipped with adapted spectro-imaging instruments, should further open up the prospects of this already booming field of research.

Another important aspect of exoplanetary exploration is the study of the atmospheric composition of exoplanets, which is also booming. In line with the work carried out by the *HST* and *Spitzer* space observatories, the *JWST* (*James Webb Space Telescope*, figure 9.5), equipped with a 6.5-m diameter mirror and optimized

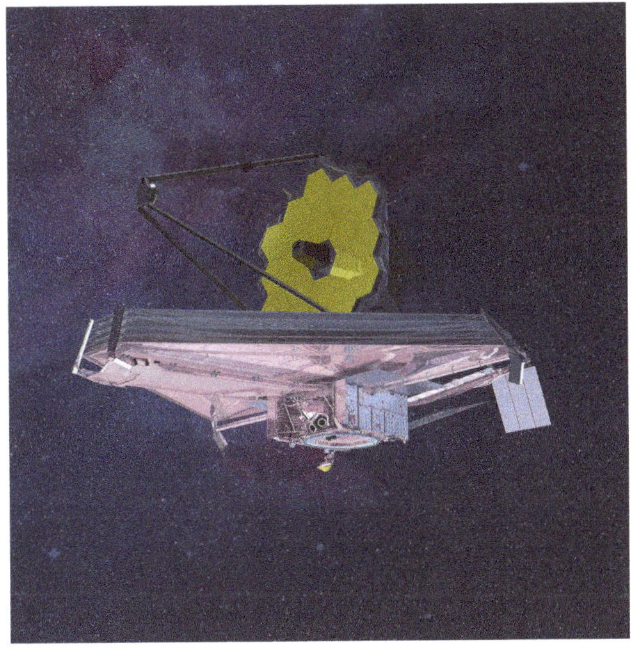

FIG. 9.5 – The *James Webb Space Telescope* (*JWST*), *Extremely Large European Telescope* (*ELT*) developed by NASA in partnership with ESA, will be the successor of the *Hubble Space Observatory* (*HST*), itself in operation since 1989. It is a 6.5-m diameter telescope optimized for observation in the visible and infrared, mainly dedicated to cosmology and the study of exoplanets. Much awaited by the entire astronomical community worldwide, it has suffered multiple delays. Its launch is currently scheduled for 2021. © NASA.

for infrared observations, will be an ideal instrument to measure the thermal infrared spectrum of exoplanets orbiting the brightest stars, in order to determine their atmospheric composition and thus reveal their nature, and even their history; its launch, long delayed and eagerly awaited by astronomers, should hopefully take place in 2021.

However the *JWST*, equipped with high spatial resolution spectrometers, will only be able to observe a limited number of exoplanets, given the diversity of its scientific objectives, ranging from stellar physics to galaxies and cosmology. A mission dedicated to the systematic analysis of the atmospheric composition of exoplanets, capable of making a census of all the types of exoplanets encountered, was therefore needed. This is the goal of the *ARIEL* (*Atmospheric Remote-sensing Infrared Exoplanet Large survey*) mission, approved by the European Space Agency for a launch in 2028. Its spectrometer will be able to simultaneously observe the infrared spectrum of the exoplanet by transit in front of its star, as well as the visible spectrum of the star in question, this to eliminate possible uncertainties related to the variability of the star or the seasonal cycle of the planet. In parallel, transit spectroscopy measurements will be performed on the most favorable targets by

ELTs, equipped with high-resolution spectrometers, in the near-infrared atmospheric windows.

To conclude this overview, let us mention two American projects that are still in the study phase, but which could concretize tomorrow's ambitions. *LUVOIR* and *HabEx* are two projects for large telescopes operating from ultraviolet to near infrared, both at the Lagrange L2 point, with the objective of imaging and characterizing exoplanets. *LUVOIR* would be equipped with a telescope with a diameter between 8 and 16 m, equipped with an adaptive optics system, a coronagraph (which could consist of a long-distance occulter), imagers and spectrographs; its scientific objectives would be multiple, ranging from the study of exoplanets to stellar and extragalactic physics, cosmology and high-energy physics. *HabEX*, equipped with a 3- to 6-m telescope also equipped with a coronagraph, would have, as its main objective, the imaging and spectroscopic characterization of habitable exoplanets, as well as the search for their possible biosignatures. Both projects are currently under study for presentation to the US National Academy of Sciences in the framework of the Decadal Survey, for possible selection in 2021 and launch around 2035.

9.6 What If We Were Not Alone?

In the light of recent discoveries, can we try to estimate the probability of the existence of extraterrestrial life? The question is not new. In 1961, the American astronomer Frank Drake went even further: his famous equation $N = F^* \times F_P \times F_H \times F_V \times F_I \times F_R \times T$ aims at evaluating the number of civilizations detectable in our Galaxy (see box 9.1). More than fifty years later, we have made great progress on the first three factors, which are the annual rate of star formation in the Galaxy, the fraction of stars surrounded by planets, and the fraction of planets in the habitable zone. These advances are encouraging, but we still do not know anything about the next terms. The first unknown is the probability that a planet located in the habitable zone of its star would be able to support life. In the case of the Earth, several factors may have been favorable to the emergence of life: the presence of the Moon stabilized the planet's axis of rotation, and the presence of a magnetosphere protecting the atmosphere and surface from the solar wind. Then, the evolution of life was marked by external events (meteorite falls), or geophysical (appearance of continents, volcanism, plate tectonics) or climatic events (periods of glaciation). In other words, life on Earth has followed its own path, specific to its environment, and it would be illusory to hope to find a form of extraterrestrial life similar to ours. If biologists agree on the conditions that seem definitely needed for the blossoming of life – liquid water, carbon, energy source, longevity – we could even imagine extraterrestrial life based not on DNA or RNA, like ours, but on other nucleic acids using amino acids not present in the list of twenty "terrestrial" amino acids.

Suppose we discover, in a few decades or perhaps less, the ozone signature in the infrared spectrum of a rocky exoplanet. Suppose that this atmosphere also shows the spectral signatures of water vapor, methane and carbon dioxide – molecules already identified on some exoplanets. The presence of ozone would mean the presence of a significant amount of oxygen. One could then seriously question the existence of some form of life on the surface of this exoplanet.

What should we do then? Impossible to go and see... On the other hand, it is possible to imagine a form of communication based on the transmission of electromagnetic waves. The idea is not new: radio astronomers, as early as the 1970s, wondered about the possibility of receiving signals emitted by possible extraterrestrial civilizations. They chose the wavelength of 21 cm, corresponding to a well-known transition of atomic hydrogen, the most abundant element in the Universe, to which the Earth's atmosphere (and also, by analogy, that of a rocky exoplanet) is transparent. A systematic research program has been launched in the direction of a few targets chosen among nearby stars, from the largest radio telescopes (figure 9.6). The searches, negative to date, were coordinated from 1984 by the SETI Institute (*Search for Extra-Terrestrial Intelligence*), first with funding from NASA, then with private funds.

In addition to listening to the "others", astronomers also emitted, in 1974, from the giant radio telescope at Arecibo, a radio message (at 12.6 cm wavelength) towards one of the globular clusters of our Galaxy located at 6400 parsecs. If an answer were to reach us, it would be in more than 40 000 years... In the same vein, the messages contained by the *Pioneer* and *Voyager* space probes, as they travel outward from the Solar system, testify to humanity's need to speak to "others". These shipments of bottles to the sea, of a purely symbolic significance, are there only to recall the existential questioning of humanity on its place in the Universe.

Box 9.1 The Drake equation. In 1961, the radio astronomer Frank Drake, born in 1930, proposed to quantify the number of extraterrestrial civilizations capable of communicating with ours. This number N is expressed by means of the following equation:

$$N = F^* \times F_P \times F_H \times F_V \times F_I \times F_R \times T$$

in which

- F^* is the annual rate of star formation in the Galaxy;
- F_P is the fraction of stars surrounded by planets;
- F_H is the number of planets per star in the habitable zone;
- F_V is the fraction of habitable planets sheltering life;
- F_I is the fraction of inhabited planets harboring civilizations;
- F_R is the fraction of these planets capable of communicating by radio;
- T is the life time of a civilization that can communicate by radio.

In 1961, the values used by Drake were as follows: $F^* = 10$, $F_P = 0.5$; $F_H = 2$, $F_V = 1$, $F_I = F_R = 0.01$, $T = 10\,000$ years, which gave $N = 10$. Recent research confirms the first two values, while the value of F_H is still uncertain. All other factors are speculative.

FIG. 9.6 – The radio telescope at Arecibo, on the island of Puerto Rico in the United States. It consisted of a huge antenna with a diameter of more than 300 m. We see here the focal carriage over the antenna. In 2020, the antenna was severely damaged after the fall of two cables and the focal carriage finally collapsed on December 1, 2020, making the end of the radio telescope's life. Wikimedia Commons.

Glossary

Accretion: gravitational capture of a solid or gaseous material by a celestial body.

Albedo: fraction of the light of the central star reflected or scattered by a planet or a comet. One distinguishes the monochromatic albedo, at a given wavelength, and the bolometric albedo or Bond albedo, which concerns the total energy integrated on all wavelengths.

Amino acid: molecule containing an acid radical COOH and an amine radical NH_2, basic constituent of proteins and peptides.

Asteroid: a small solid body belonging to a planetary system; most asteroids of the Solar System are located on orbits between Mars and Jupiter.

Astrometry: measurement of the position, the proper motions and the distance of celestial objects (planets, satellites, stars).

Astronomical unit: unit of length equal to the semi-major axis of the Earth's orbit (150 million km); official symbol: au.

Brown dwarf: an aborted star, whose mass is insufficient for the nuclear reactions to have been able to start and provide energy. The mass of brown dwarfs is approximately between 0.01 and 0.07 solar mass.

Carbonaceous chondrite: type of meteorite whose chemical composition is close to that of the interstellar medium as far as refractory elements are concerned.

Coma: cloud of gas and dust surrounding the nucleus of a comet.

Comet: an object belonging to a planetary system, whose orbit is generally very eccentric and very disturbed by the giant planets. Comets are solid blocks of ice and dark organic matter that evaporate superficially as they approach the Sun, releasing gas and dust responsible for their nebulous appearance and tails.

Continuous spectrum, or continuum: emission or absorption covering a wide range of wavelengths without any preferred wavelength.

Coronagraph: an instrument that occults very effectively the light of a star in order to observe weak objects in the vicinity.

Cosmogony: in the general sense, study of the formation and evolution of celestial bodies. This word is usually taken in a restrictive sense and applies only to the Solar System.

Cryovolcanism: type of volcanism observed in certain icy objects, where molten products, generally under the effect of tidal forces, are ejected in the form of geysers of water, ammonia or methane.

Cyanobacteria: a type of prokaryotic bacteria producing oxygen from CO_2.

Differentiation: gravitational sorting of elements according to their mass within a star or another astronomical object.

Doppler–Fizeau (effect): variation $\Delta\lambda$ of the wavelength λ received from a source in motion, with respect to the wavelength λ_0 emitted at rest. We have $\Delta\lambda/\lambda_0 = v/c$, where v is the radial velocity of the source and c the velocity of light.

Eccentricity: for an orbit, deviation from circularity. The eccentricity e of an elliptical orbit is such that the distance from the center to one of the foci of the ellipse is ae, where a is the semi-major axis of the orbit.

Eclipse: see *transit*.

Ecliptic: apparent trajectory of the Sun among the stars during the year.

Equator: for a planet or a satellite, a large circle perpendicular to its rotation axis, from which the latitude is counted. For the sky, the celestial equator is the projection on the celestial sphere of the equator of the Earth.

Eukaryote: living cell with a nucleus, isolated or forming fungi, plants and animals.

Exobiology: study of living matter outside the Earth, or more modestly, for the time being, of its precursors.

Exoplanet: planet gravitating around another star than the Sun.

Giant planet: planet with a mass much higher than 10 Earth masses, generally very voluminous and not very dense. One distinguishes in the Solar System the gaseous giant planets (Jupiter and Saturn) and the icy giant planets (Uranus and Neptune).

Habitability zone: range of distances to a star such that water can be in a liquid state on a possible exoplanet.

Ice line: circle of the circumstellar disk such that water is in the vapor form inside and in the solid form beyond it.

Inclination: angle between the plane of a planetary orbit and a reference plane, which for the Solar System is the plane of the ecliptic. For the orbit of an exoplanet, the plane of reference is the plane of the sky.

Interstellar matter: gas and dust filling the Milky Way between the stars. It is the material from which stars and planets form.

Kuiper belt: region of the Solar System where icy objects like Pluto are found, located between 30 and 55 astronomical units of the Sun.

Late Heavy Bombardment: intense bombardment of the planets by asteroids, occurring about 880 million years after the formation of the Solar System.

Lipids (fats): molecules formed of long carbon chains with usually a phosphoglyceride at one end. They are the constituents of the walls of living cells.

Lithosphere: solid crust of a rocky planet.

Magnetosphere: the upper region of the atmosphere of a planet with a magnetic field, such as the Earth, Jupiter and Saturn, characterized by the presence of high-energy charged particles trapped in this magnetic field.

Magnitude: a logarithmic scale for measuring the brightness of a star;

– *apparent magnitude*: measures the apparent brightness: $m = -2.5 \log F +$ constant, where F is the luminous flux;
– *absolute magnitude*: measures the intrinsic brightness. By convention, the absolute magnitude M and the apparent magnitude m of a star would be identical if it was at a distance of 10 parsecs (32.6 light years): $m - M = 5 - 5 \log D$, D being the distance of the object in parsecs.

Meteorite: debris of a solid body belonging to the Solar System, whose trajectory crossed that of the Earth and entered its atmosphere. The luminous phenomenon associated with this crossing of the Earth's atmosphere is a shooting star, the residue (if there is one) is a meteorite. Meteorites are very primitive rocks; carbonaceous chondrites are the most interesting for understanding the formation of the Solar System.

Methanogenic bacteria: anaerobic prokaryotic bacteria, classified as archaea, producing methane from H_2 and CO_2.

Migration: radial displacement of young planets due to their gravitational interaction with the protoplanetary disc.

Nebula: any astronomical object of diffuse aspect, except comets. *Planetary nebulae*, which have long been thought to be at the origin of planetary systems, are actually masses of gas and dust expelled by end-of-life stars.

– *Protosolar nebula*: mass of interstellar gas and dust from which the Solar System was formed.

Nucleic acid: characteristic macromolecule of the terrestrial living matter, consisting of a chain of nucleotides joined by phosphate groups. Ribonucleic acid (RNA) is a single chain, while deoxyribonucleic acid (DNA), the basic constituent of chromosomes, is in the form of a double helix.

Nucleotide: molecule formed of a sugar base and nitrogen rings. Deoxyribonucleic acid (DNA) and ribonucleic acid consist of nucleotides linked together by phosphate groups. The order in which the nucleotides are arranged in the DNA defines the genetic code.

Occultation: see *transit*.

Oort's cloud: area located at the edge of the Solar System, about 1 light year of the Sun, where the comets reside until a gravitational disturbance eventually sends them near the Sun.

Panspermia: the theory that life would have been brought from outside on Earth. In a mitigated form, the theory that only the basic building blocks of life have been brought.

Peptide: molecule formed from a combination of amino acids.

Perihelion or periaster: point in the orbit of a planet where it is closest to the central star.

Planetesimal: a small body formed by agglomeration of dust in the protoplanetary disk.

Planetoid: a larger body formed by accretion of planetesimals (see this word) following their mutual collisions; the bigger ones accrete gas and become planets, in the model of Safronov.

Plate tectonics: relative displacement of rigid portions of the Earth's lithosphere (continental plates). The other planets of the Solar System do not have this property.

Prebiotic molecule: an organic molecule, that is to say formed essentially of the atoms H, C, N and O, which could be a basic brick of the proteins and the DNA. Amino acids are such molecules.

Prokaryote: a living cell, bacteria or archaea, without a nucleus.

Protoplanetary disk: disk of gas and dust rotating around a proto-star, in which the planets will form. When the planets are formed, they affect the structure of the disk, which is then called a *transition disk*. Finally, when all the gas of the disk has disappeared, a *debris disk* remains.

Pulsar: neutron star in very fast rotation, regularly emitting, like a beacon, a radio signal whose period is extremely stable.

Radial velocity: the velocity of approach or recession of an object, measured by the Doppler–Fizeau effect on its spectral lines.

Radio astronomy: branch of astronomy which consists in studying the radio emission of celestial objects. The Sun, planets, certain stars, the atomic, molecular or ionized interstellar gas, high-energy electrons of the cosmic radiation, pulsars, galaxies and quasars emit radio waves.

Radio telescope: antenna or set of antennas used for radio astronomy.

Revolution: movement of a planet around its star, or a satellite around its planet.

Rotation: movement of a planet or a satellite around its axis.

Solar wind: ionized gas continuously ejected by the Sun's corona.

Spectral band: a set of often unresolved spectral lines that produce absorption or emission covering an extended range of wavelengths. The bands are characteristic of the molecules, which produce very many lines.

Spectral line: increase or decrease of intensity in the spectrum of an object occurring at a given wavelength; the line is called emission if there is reinforcement, and absorption if there is decrease. The wavelength of a line is characteristic of the atom, ion or molecule which produces it.

Spectroscopy: technique consisting in decomposing by a prism or a diffraction grating the light in its different wavelengths. By extension, decomposition of any electromagnetic wave (ultraviolet, infrared or radio). The instrument used is called *spectroscope* if it is not recording, or *spectrograph* if it is. Spectroscopy can be *in emission* if one observes a luminous body as an exoplanet, or *in transmission* if one observes for example the atmosphere of an exoplanet interposed in front of a star.

Supernova: massive star ending its life in an explosion.

Super-Earth: rocky exoplanet with a mass of about ten Earth masses.

Telluric (or terrestrial) planet: rocky planet in the Solar System comparable to the Earth: Mercury, Venus, Earth and Mars, as opposed to the giant planets (Jupiter, Saturn, Uranus and Neptune).

Tide: deformation or rupture of an object under the effect of the gravitation of a neighboring star.

Transit: passage of a planet in front of its star (*primary transit*) producing an eclipse of the star by the planet, or passage behind the star (*secondary transit*), producing an occultation of the planet by the star.

Velocimetry: method of indirect detection of an exoplanet by the variations of radial velocity that it produces on its star.

Bibliography

Bougher S.W., Hunten D.M., Phillips R.J. (1997) *Venus II: geology, geophysics, atmosphere and solar wind environment.* University of Arizona Press, Tucson.

Catling D.C., Kasting J.F. (2017) *Atmospheric evolution of inhabited and lifeless worlds.* Cambridge University Press, Cambridge.

Connes P. (2020) *History of the plurality of worlds* (J. Lequeux, Ed.). Springer Nature Switzerland AG.

Coustenis A., Encrenaz T. (2013) *Life beyond Earth.* Cambridge University Press, Cambridge.

Coustenis A., Crovisier J., Fulchignoni M., Lamy L., Roques F., Zanda B. (2021) *The solar system, origin and evolution* (T. Encrenaz, J. Lequeux, Eds), 2 vol. ISTE, London.

Encrenaz T. (2013) *Planets, ours and others – from Earth to exoplanets.* EDP Sciences and World Scientific.

Forget F., Costard F., Lognonné F. (2007) *Planet Mars, story of another world.* Springer, New York.

Haldane J.B.S. (1985) *On being the right size and other essays.* Oxford University Press, Oxford.

Kieffer H.H., Jakosky B.M., Snyder C.W., Matthews M.S. (1992) *Mars.* University of Arizona Press, Tucson.

Lequeux J., Encrenaz T., Casoli F. (2020) *The exoplanet revolution.* EDP Sciences, Les Ulis.

Lissauer J.J., de Pater I. (2019) *Fundamental planetary science – physics, chemistry and habitability,* 3rd edn. Cambridge University Press, Cambridge.

McSween H.Y., Jr. (1999) *Meteorites and their parent planets.* Cambridge University Press, Cambridge.

Ollivier M., Encrenaz T., Roques F., Selsis F., Casoli F. (2009) *Planetary systems: detection, formation and habitability of extrasolar planets.* Springer, Berlin.

Perryman M. (2018) *The exoplanet handbook,* 2nd edn. Cambridge University Press, Cambridge.

Rothery D.A., Gilmour I., Sephton M.A. (2018) *An introduction to astrobiology,* 3rd edn. Cambridge University Press, Cambridge.

Schulze-Makuch D., Bains W. (2017) *The cosmic zoo, complex life on many worlds.* Springer, New-York.

Schulze-Makuch D., Irwin L.N. (2018) *Life in the universe.* Springer Praxis Book, Springer Nature Switzerland AG.

Taylor F.W. (2014) *The scientific exploration of Venus.* Cambridge University Press, Cambridge.

By Internet

Updated list of exoplanets, with data and bibliography: http://exoplanet.eu

Impression & brochage - France
Numéro d'impression : N04545210409 - Achevé d'imprimer : Mai 2021
Dépôt légal : Mai 2021

PEFC 10-31-3532 / **Certifié PEFC** / Ce produit est issu de forêts gérées durablement et de sources contrôlées. / pefc-france.org